Frontiers in Antimicrobial Agents

(Volume 1)

The Challenging of Antibiotic Resistance in the Development of New Therapeutics

Editors

Manuela Oliveira & Isa Serrano

Centro de Investigação Interdisciplinar em Sanidade Animal,

Faculdade de Medicina Veterinária,

Universidade de Lisboa, Avenida da Universidade Técnica,

1300-477 Lisboa,

Portugal

Bentham Science Publishers Ltd.
Executive Suite Y - 2
PO Box 7917, Saif Zone
Sharjah, U.A.E.
Email: subscriptions@benthamscience.org

BENTHAM SCIENCE

CONTENTS

The Cover Image has been provided by Dr. Manuela Oliveira, the Editor of this book.

The Challenging of Antibiotic Resistance in the Development of New Therapeutics

FOREWORD

The rapid development of highly effective antimicrobial agents during the 20th century revolutionized the treatment of diseases caused by bacteria, viruses and fungi; leading to the notion that "it was time to close the book on infectious diseases". However, today we are facing pandrug-resistant microorganisms, and antimicrobial resistance constitutes one of the major public health problems worldwide. Allied to this, the new antimicrobial agent's development pipeline is at its all-time low; because of scientific, economic and regulatory hurdles.

While we must conserve the antimicrobials we have left by using them optimally, the process of developing new agents must also be accelerated. This will hopefully facilitate targeted therapy, improving therapeutic efficacy and decreasing antimicrobial resistance.

This book describes cutting-edge research on innovative alternatives to classical antimicrobial therapy – bacteriophages, antimicrobial peptides, probiotics, immunomodulators, natural compounds, bacteriocins and biocides – and the most appropriate approaches to control the spread of drug-resistant microorganisms.

Interestingly, this book is edited and partly written by scholars dedicated to Microbiology from the area of Veterinary Medicine. This is not surprising, because the amount of antimicrobials marketed for use in animals is approximately four times greater than the quantity used in human medicine. Furthermore, the widespread use of antimicrobial agents in animal production – often administered in lower doses and for longer periods of time – has been linked to the development of antimicrobial resistance.

While the novel therapeutic strategies in Veterinary Medicine have been a major focus in the last chapter, the book has input from a wide range of experts in different disciplines – from basic science to human clinical microbiology – and truly reflects the 'One-Health' approach which spans humans, animals and the wider environment.

One final note to remember by the enthusiastic reader is that, bacteria have shown, in this continuous "arms race", that they can develop resistance to virtually all therapeutic agents. Therefore, it is very important to continue to use both antibiotics and their alternatives rationally and judiciously.

João João Mendes
Emergency Medicine and Intensive Care Internal Medicine Department,
Santa Marta's Hospital/Central Lisbon Hospital Center,
Lisbon, Portugal

PREFACE

This eBook was conceived after an invitation by the Bentham Science Publishers due to the necessity to review the alternative approaches to the classic antibiotic therapies, and to disperse the shadows over the use of these innovative medicines.

The prevalence of drug-resistant microorganisms has increased worldwide. However, in the last years, few novel entities to fight drug resistant microorganisms have entered the clinic. The antibacterial pipeline is scarce because of the costs associated with the clinical trials and licensing of the antibiotics. Therefore, the quest for alternatives to the classic antimicrobial therapeutics that are effective against drug resistant microorganisms is a timely and very important issue in modern medicine.

The development of novel entities to control the dissemination of drug resistant microorganisms would decrease the morbidity and mortality caused by these disease causing agents, and consequently the economic burden associated with health treatments. Some of these drugs are considered extremely valuable in the nearly future due to their low toxicity, capacity for large scale production, and most importantly their low probability to generate resistance. However, the fear associated to the use of innovative medicines such as phage therapy must be overcome. It will take time to bring phage therapy and other approaches to practice with safety, and to change mentalities of clinicians and the general public. The implementation of such innovative medicines will lead to a decrease in the antimicrobial resistance and related failure treatments. This progress will require contribution of different levels of interdisciplinary knowledge: from researchers to public health entities, from producers to consumers, including politicians.

This eBook aims to contribute to an integrated understanding concerning innovative alternatives to the classical antimicrobial therapeutics. It is based on cutting edge research and the outcome will shed light on the most appropriate approach to control the dissemination of drug resistant microorganisms. It gathers a wide range of topics on the subject and includes several chapters with original material. Authors and co-authors represent a multidisciplinary team that includes scientists and professors with a vast experience in the area, from different universities and research institutions. It is an attempt to encourage the implementation of alternative approaches to the classic antimicrobial therapeutics in human and veterinary health programmes.

The eBook is organized in nine chapters: the general introduction is followed by a review devoted to Phage Therapy, Antimicrobial Peptides (AMPs), Probiotics, Immunotherapy, Natural Compounds, Bacteriocins, Biocides, and lastly to the Novel Therapeutic Strategies in

Veterinary Medicine.

Finally, we would like to thank all authors that have enthusiastically contributed to this eBook, and all people that somehow helped us to bring it to daylight, including our family, friends, colleagues and students.

Manuela Oliveira & Isa Serrano
Centro de Investigação Interdisciplinar em Sanidade Animal
Faculdade de Medicina Veterinária
Universidade de Lisboa
Avenida da Universidade Técnica
Lisboa
Portugal

List of Contributors

Amélia Fernandes	Regulatory Affairs Portugal, Ceva Saúde Animal, Rua Dr. António Loureiro Borges, 9/9A - 9ºA, Miraflores, 1495-131 Algès, Portugal
Ana Patrícia Quendera	Centro de Investigação Interdisciplinar em Sanidade Animal, Faculdade de Medicina Veterinária, Universidade de Lisboa, Avenida da Universidade Técnica, Lisboa, Portugal
Antonio Gálvez Ruiz	Departamento de Ciencias de la Salud, Facultad de Ciencias Experimentales, Universidad de Jaén, Campus Las Lagunillas, Jaén, 23071, Spain
Ana P. Duarte	Centro de Investigação Interdisciplinar em Sanidade Animal (CIISA), Faculdade de Medicina Veterinária da Universidade de Lisboa, Avenida da Universidade Técnica, Lisboa, 1300-477, Portugal
Carla Mottola	CIISA/Faculty of Veterinary Medicine of the University of Lisbon, Avenida da Universidade Técnica, 1300-477 Lisboa, Portugal
Daniel De Vos	Laboratory for Molecular and Cellular Technology (LabMCT), Queen Astrid Military Hospital, Brussels, Belgium
Dias J.	Centro de Investigação Interdisciplinar em Sanidade Animal (CIISA), Faculdade de Medicina Veterinária, Technical University of Lisbon (TULisbon), Av. Universidade Tècnica, Lisbon, Portugal
Gilbert Verbeken	Laboratory for Molecular and Cellular Technology, Queen Astrid Military Hospital, Brussel, Bruynstraat 1, B-1120, Belgium
Hikmate Abriouel	Area de Microbiología, Departamento de Ciencias de la Salud, Facultad de Ciencias Experimentales, Universidad de Jaén, Jaén, Spain
Isa Domingues Serrano	Centro de Investigação Interdisciplinar em Sanidade Animal (CIISA), Faculdade de Medicina Veterinária da Universidade de Lisboa, Avenida da Universidade Técnica, Lisboa, 1300-477, Portugal
Jean-Paul Pirnay	Laboratory for Molecular and Cellular Technology, Queen Astrid Military Hospital, Brussel, Bruynstraat 1, B-1120, Belgium
Leal R.	Centre Hospitalier Vétérinaire Fregis, 43 Av. Aristide Briand 94110, Arcueil, France
Leyre Lavilla Lerma	Area de Microbiología, Departamento de Ciencias de la Salud, Facultad de Ciencias Experimentales, Universidad de Jaén, Jaén, Spain
Lourenço A.M.	Centro de Investigação Interdisciplinar em Sanidade Animal (CIISA), Faculdade de Medicina Veterinária, Technical University of Lisbon (TULisbon), Av. Universidade Tècnica, Lisbon, Portugal
Mario Vaneechoutte	Laboratory for Bacteriology Research, Faculty of Medicine, Ghent University, Ghent, Belgium

María del Carmen Casado Muñoz	Area de Microbiología, Departamento de Ciencias de la Salud, Facultad de Ciencias Experimentales, Universidad de Jaén, Jaén, Spain
Manuela Oliveira	Centro de Investigação Interdisciplinar em Sanidade Animal (CIISA), Faculdade de Medicina Veterinária da Universidade de Lisboa, Avenida da Universidade Técnica, Lisboa, 1300-477, Portugal
Maya Merabishvili	Laboratory for Molecular and Cellular Technology, Queen Astrid Military Hospital, Brussel, Bruynstraat 1, B-1120, Belgium Laboratory for Bacteriology Research, Faculty of Medicine, Ghent University, Ghent, Belgium Microbiology and Virology, George Eliava Institute of Bacteriophages, Tbilisi, Georgia
Nabil Benomar	Area de Microbiología, Departamento de Ciencias de la Salud, Facultad de Ciencias Experimentales, Universidad de Jaén, Jaén, Spain
Rubén Pérez Pulido	Area de Microbiología, Departamento de Ciencias de la Salud, Facultad de Ciencias Experimentales, Universidad de Jaén, Jaén, Spain
Rui Seixas	Centro de Investigação Interdisciplinar em Sanidade Animal (CIISA), Faculdade de Medicina Veterinária da Universidade de Lisboa, Avenida da Universidade Técnica, Lisboa, 1300-477, Portugal
Sara Correia Santos	Eco-evolutionary Genetics Group, Instituto Gulbenkian de Ciência, Rua da Quinta Grande 6, Oeiras, Portugal
Serge Jennes	Burn Wound Centre, Queen Astrid Military Hospital, Bruynstraat 1, B-1120, Brussel, Belgium
Solange Gil	Centro de Investigação Interdisciplinar em Sanidade Animal (CIISA), Faculdade de Medicina Veterinária da Universidade de Lisboa, Avenida da Universidade Técnica, Lisboa, 1300-477, Portugal
Tavares L.	Centro de Investigação Interdisciplinar em Sanidade Animal (CIISA), Faculdade de Medicina Veterinária, Technical University of Lisbon (TULisbon), Av. Universidade Tècnica, Lisbon, Portugal
Teresa Semedo-Lemsaddek	Centro de Investigação Interdisciplinar em Sanidade Animal (CIISA), Faculdade de Medicina Veterinária da Universidade de Lisboa, Avenida da Universidade Técnica, Lisboa, 1300-477, Portugal

CHAPTER 1

Introduction

Manuela Oliveira[1,*], Amélia Fernandes[2], Isa Serrano[1]

[1] *Centro de Investigação Interdisciplinar em Sanidade Animal, Faculdade de Medicina Veterinária, Universidade de Lisboa, Avenida da Universidade Técnica, 1300-477 Lisboa, Portugal*

[2] *Regulatory Affairs Portugal, Ceva Saúde Animal, Rua Dr. António Loureiro Borges, 9/9A - 9ºA, Miraflores, 1495-131 Algès, Portugal*

Abstract: Antibiotherapy protocols for infectious disease control were first applied in 1940s with penicillin. Although they led to a major decrease in morbidity and mortality rates, they were also responsible for the rapid emergence and dissemination of multi-drug-resistant bacteria (MDR). This phenomena led to the prompt development of new antimicrobial compounds, which soon became ineffective due to bacteria ability to develop resistant traits through mutations or resistance genes transfer.

However, the development of new approaches for prevention and control of emerging infections remains one of the major priorities and challenges for Research and Innovation (R&I). The worldwide mortality rate due to infectious diseases keeps exponentially increasing, not only due to MDR bacteria dissemination, but also due to the decline in the development and commercialization of new generations of antibacterial compounds.

Considering the "One Health" concept, new antibacterial strategies are urgent, for human and veterinary medicine. Several R&I approaches are being followed, including phage therapy, antimicrobial peptides and bacteriocins, probiotics, natural compounds, immunomodulation *via* vaccination and biocides.

Keywords: Antimicrobial peptides, Antimicrobial resistance, Bacteriocins, Bacteriophages, Biocides, Biofilm, Generally Recognized as Safe, Host, Immunomodulation, Lytic, Multi-Drug-Resistant, Natural compounds, Nisin, "One Health", Pathogens, Probiotics, Research and innovation, Therapeutic strategies, Vaccination, Veterinary Medicine.

* **Corresponding author Manuela Oliveira:** Centro de Investigação Interdisciplinar em Sanidade Animal (CIISA), Faculdade de Medicina Veterinária da Universidade de Lisboa, Avenida da Universidade Técnica, 1300-477, Lisboa, Portugal; E-mail: moliveira@fmv.ulisboa.pt

Manuela Oliveira and Isa Serrano (Eds)

INTRODUCTION

Antimicrobial resistant bacteria dissemination poses an emergent challenge for the development of new antimicrobial compounds [1], which has significantly diminished in the last decade. Therapeutic options to control the dissemination of resistant bacteria are progressively decreasing. Considering the "One Health" concept, new antibacterial strategies are urgent for human and veterinary medicine. Novel therapeutics strategies aiming at infectious diseases control include phage therapy, probiotics, immunomodulation *via* vaccination, natural compounds and biocides and bacteriocins and antimicrobial peptides.

Antimicrobial peptides (AMP) may constitute new potential antimicrobial compounds: they are able of eliminating a broad spectrum of microorganisms; resistance to AMP was rarely observed; and they have the ability to modulate innate immune response [2]. AMPs are a diverse group of molecules with cationic and amphipathic properties which selectively target membranes rich in negatively charged phospholipids and no cholesterol, such as those of microorganisms [2, 3]. Most have amino acids with two types of side chains: cationic, such as arginine (R), lysine (K), and histidine (H); and bulky nonpolar, *e.g.*, proline (P), phenylalanine (F), and tryptophan (W). The former presumably mediates interactions with bacterial membranes and/or cell walls, including lipo-polysaccharide [4], whereas nonpolar chain mediates lipophilic attachment leading to membrane disruption [2, 3].

Recently, it was established that short AMPs having sequence (RW) n-NH2, (n= 2 to 4) are the optimal choice against bacteria. Hexameric and octameric peptides are potent biofilm inhibitors and octameric peptides also disperse existing biofilms and kill detached cells [3, 4]. Other advantage of short AMPs is their minimum damage to host cells, being candidates to large-scale production [3]. In addition, it was shown that some substances that disturb biofilm matrix make them more susceptible to subsequent treatment with antimicrobials or disinfectants, namely furanone [5], nisin [6], and salicylic acid [6]. Cellulase also inhibits biofilm formation by Gram-negative bacteria, probably by degrading the exopolyssacharide [7].

Bacteriocins are an heterogeneous group of ribosomally synthesized AMP produced by bacteria, divided in several classes [8]. Nisin is a class I bacteriocin produced by *Lactococcus lactis* that has antibacterial activity against a wide range of Gram-positive bacteria [9 - 11]. Nisin is already approved as a food preservative in EU (as additive E234) and USA [9].

Some bacteriocins are produced by probiotic strains. Probiotics are defined as "live microorganisms whose when administered in adequate amounts confer a health benefit on the host". Probiotics are documented to reduce or prevent specific infectious diseases of the gastrointestinal tract. Proposed mechanisms by which probiotics can inhibit pathogens in the gut environment include (i) competitive exclusion by competition for binding sites or stimulation of epithelial barrier function; (ii) stimulation of immune responses *via* increases of sIgA and anti-inflammatory cytokines and regulation of proinflammatory cytokines; (iii) direct antimicrobial activity through production of antimicrobial substances, including bacteriocins and other antimicrobial peptides (AMPs); (iv) production of biosurfactant molecules with antiadhesive and sometimes also antimicrobial activity; (v) inhibition of key cellular processes in gastrointestinal pathogens such as virulence gene or protein expression, quorum sensing or biofilm formation [12, 13].

Bacteriophages are ubiquitous bacterial viruses [14 - 20] that specifically invade several bacterial genera [21, 22]. They are obligate parasites which nucleic acid codes for the proteins necessary for its replication within the host bacterium [14] and can be lytic or temperate [14, 16, 20 - 22]. Bacteriophages used in therapy are usually lytic. Their life cycle includes adsorption to a specific phage receptor on the bacterial cell surface, genomic material injection, exponential phage replication using the host cell machinery, bacteria lysis and release of progeny phages [14, 16, 18, 21, 22]. These phages can present two bacteriolytic mechanisms, promoting cell wall hydrolysis *via* a virolysin-holin system present in most lytic phages, or *via* a single lytic factor [20].

One main advantage of phages is their host specificity [15 - 18, 20, 22], which renders them ideal for eliminating only pathogenic bacteria, belonging to a specific species, serovar or serogroup [14, 16, 20, 22]. Unrelated commensal microbiota is not affected by phage presence [16, 20, 21], which constitutes a major advantage regarding antibiotic treatment [16, 21]. Other advantages include the fact that phages are able to eliminate antimicrobial-resistant bacteria [17, 18, 20]; they are cost effective and easy to produce [15, 20 - 22]; they can be combined with antimicrobial compounds or other phages for a synergic effect [22]; and they are safe [15, 17, 18]. Some phage products have already been approved as Generally Recognized As Safe (GRAS) by the Food and Drug Administration [15, 20, 22, 23].

Many natural compounds are also already approved as GRAS, namely several plant extracts. These compounds have been studied regarding their potential to be used as an effective alternative to current antibiotics and disinfectants, and the antibacterial properties of plants such as basil, cinnamon, clove, mint, oregano, salvia and thyme have already been long described [24 - 27]. They are also promising compounds for prevention and treatment of infection promoted by MDR bacteria, as: there are no evidence regarding the relation of these compounds with antimicrobial resistance dissemination; they are usually not toxic for eukaryotic cells; and they are eco-friendly, being easily biodegraded [28].

Amongst Research and Innovation (R&I) approaches that are being followed, immunomodulation including vaccines and bacterins, seems extremely promising, as it stimulates an organic system already involved in infections control, and not by distressing the bacterial agent itself, a mechanism often related with resistance development not being related to resistance dissemination [29].

To finalize, it is of major importance to refer that all these new R&I approaches, including phage therapy, antimicrobial peptides and bacteriocins, probiotics, natural compounds, immunomodulation *via* vaccination and biocides, can also be applied to veterinary medicine. Considering the "One Health" concept, this would be of major importance, to avoid and control the dissemination of MDR strains between animals, humans and the environment [30, 31].

CONFLICT OF INTEREST

The authors confirm that this chapter contents have no conflict of interest.

ACKNOWLEDGEMENTS

Declared none.

REFERENCES

[1] Bassetti M, Ginocchio F, Mikulska M, Taramasso L, Giacobbe DR. Will new antimicrobials overcome resistance among Gram-negatives? Expert Rev Anti Infect Ther 2011; 9(10): 909-22. [http://dx.doi.org/10.1586/eri.11.107] [PMID: 21973303]

[2] Baltzer SA, Brown MH. Antimicrobial peptides: promising alternatives to conventional antibiotics. J Mol Microbiol Biotechnol 2011; 20(4): 228-35. [http://dx.doi.org/10.1159/000331009] [PMID: 21894027]

[3] Liu Z, Brady A, Young A, *et al.* Length effects in antimicrobial peptides of the (RW)n series. Antimicrob Agents Chemother 2007; 51(2): 597-603. [http://dx.doi.org/10.1128/AAC.00828-06] [PMID: 17145799]

[4] Hou S, Liu Z, Young AW, Mark SL, Kallenbach NR, Ren D. Effects of Trp- and Arg-containing antimicrobial-peptide structure on inhibition of Escherichia coli planktonic growth and biofilm formation. Appl Environ Microbiol 2010; 76(6): 1967-74. [http://dx.doi.org/10.1128/AEM.02321-09] [PMID: 20097816]

[5] Vestby LK, Lönn-Stensrud J, Møretrø T, *et al.* A synthetic furanone potentiates the effect of disinfectants on Salmonella in biofilm. J Appl Microbiol 2010; 108(3): 771-8. [http://dx.doi.org/10.1111/j.1365-2672.2009.04495.x] [PMID: 19735326]

[6] Rosenberg LE, Carbone AL, Römling U, Uhrich KE, Chikindas ML. Salicylic acid-based poly(anhydride esters) for control of biofilm formation in Salmonella enterica serovar Typhimurium. Lett Appl Microbiol 2008; 46(5): 593-9. [http://dx.doi.org/10.1111/j.1472-765X.2008.02356.x] [PMID: 18373656]

[7] Loiselle M, Anderson KW. The use of cellulase in inhibiting biofilm formation from organisms commonly found on medical implants. Biofouling 2003; 19(2): 77-85. [http://dx.doi.org/10.1080/0892701021000030142] [PMID: 14618691]

[8] Chopra L, Singh G, Choudhary V, Sahoo DK. Sonorensin: an antimicrobial peptide, belonging to the heterocycloanthracin subfamily of bacteriocins, from a new marine isolate, Bacillus sonorensis MT93. Appl Environ Microbiol 2014; 80(10): 2981-90. [http://dx.doi.org/10.1128/AEM.04259-13] [PMID: 24610839]

[9] Field D, Gaudin N, Lyons F, *et al.* A bioengineered nisin derivative to control biofilms of Staphylococcus pseudintermedius. PLoS One 2015; 10(3): e0119684. [http://dx.doi.org/10.1371/journal.pone.0119684] [PMID: 25789988]

[10] Schulz D, Pereira M, Bonelli R, *et al.* Bacteriocinas: mecanismo de ação e uso na conservação de alimentos. Alim Nutr 2003; 14(2): 229-35.

[11] Todorov SD, Favaro L, Gibbs P, Vaz-Velho M. Enterococcus faecium isolated from Lombo, a Portuguese traditional meat product: characterisation of antibacterial compounds and factors affecting bacteriocin production. Benef Microbes 2012; 3(4): 319-30.

[http://dx.doi.org/10.3920/BM2012.0036] [PMID: 23234731]

[12] Corr SC, Hill C, Gahan CG. Understanding the mechanisms by which probiotics inhibit gastrointestinal pathogens. Adv Food Nutr Res 2009; 56: 1-15. [http://dx.doi.org/10.1016/S1043-4526(08)00601-3] [PMID: 19389605]

[13] Kanmani P, Satish Kumar R, Yuvaraj N, Paari KA, Pattukumar V, Arul V. Probiotics and its functionally valuable products-a review. Crit Rev Food Sci Nutr 2013; 53(6): 641-58. [http://dx.doi.org/10.1080/10408398.2011.553752] [PMID: 23627505]

[14] Donlan RM. Preventing biofilms of clinically relevant organisms using bacteriophage. Trends Microbiol 2009; 17(2): 66-72. [http://dx.doi.org/10.1016/j.tim.2008.11.002] [PMID: 19162482]

[15] Górski A, Targońska M, Borysowski J, Weber-Dabrowska B. The potential of phage therapy in bacterial infections of the eye. Ophthalmologica 2009; 223(3): 162-5. [http://dx.doi.org/10.1159/000193293] [PMID: 19155637]

[16] Johnson RP, Gyles CL, Huff WE, *et al.* Bacteriophages for prophylaxis and therapy in cattle, poultry and pigs. Anim Health Res Rev 2008; 9(2): 201-15. [http://dx.doi.org/10.1017/S1466252308001576] [PMID: 19102791]

[17] Ma Y, Pacan JC, Wang Q, *et al.* Microencapsulation of bacteriophage felix O1 into chitosan-alginate microspheres for oral delivery. Appl Environ Microbiol 2008; 74(15): 4799-805. [http://dx.doi.org/10.1128/AEM.00246-08] [PMID: 18515488]

[18] Mann NH. The potential of phages to prevent MRSA infections. Res Microbiol 2008; 159(5): 400-5. [http://dx.doi.org/10.1016/j.resmic.2008.04.003] [PMID: 18541414]

[19] O'Flaherty S, Ross RP, Flynn J, Meaney WJ, Fitzgerald GF, Coffey A. Isolation and characterization of two anti-staphylococcal bacteriophages specific for pathogenic Staphylococcus aureus associated with bovine infections. Lett Appl Microbiol 2005; 41(6): 482-6. [http://dx.doi.org/10.1111/j.1472-765X.2005.01781.x] [PMID: 16305674]

[20] Parisien A, Allain B, Zhang J, Mandeville R, Lan CQ. Novel alternatives to antibiotics: bacteriophages, bacterial cell wall hydrolases, and antimicrobial peptides. J Appl Microbiol 2008; 104(1): 1-13. [PMID: 18171378]

[21] Mattey M, Spencer J. Bacteriophage therapy--cooked goose or phoenix rising? Curr Opin Biotechnol 2008; 19(6): 608-12. [http://dx.doi.org/10.1016/j.copbio.2008.09.001] [PMID: 18926909]

[22] O'Flaherty S, Ross RP, Coffey A. Bacteriophage and their lysins for elimination of infectious bacteria. FEMS Microbiol Rev 2009; 33(4): 801-19. [http://dx.doi.org/10.1111/j.1574-6976.2009.00176.x] [PMID: 19416364]

[23] Sillankorva S, Neubauer P, Azeredo J. Pseudomonas fluorescens biofilms subjected to phage phiIBB-PF7A. BMC Biotechnol 2008; 8: 79. [http://dx.doi.org/10.1186/1472-6750-8-79] [PMID: 18954451]

[24] Burt S. Essential oils: their antibacterial properties and potential applications in foods--a review. Int J Food Microbiol 2004; 94(3): 223-53.

[http://dx.doi.org/10.1016/j.ijfoodmicro.2004.03.022] [PMID: 15246235]

[25] Kon KV, Rai MK. Plant essential oils and their constituents in coping with multidrug-resistant bacteria. Expert Rev Anti Infect Ther 2012; 10(7): 775-90.
[http://dx.doi.org/10.1586/eri.12.57] [PMID: 22943401]

[26] Ríos JL, Recio MC. Medicinal plants and antimicrobial activity. J Ethnopharmacol 2005; 100(1-2): 80-4.
[http://dx.doi.org/10.1016/j.jep.2005.04.025] [PMID: 15964727]

[27] Seow YX, Yeo CR, Chung HL, Yuk HG. Plant essential oils as active antimicrobial agents. Crit Rev Food Sci Nutr 2014; 54(5): 625-44.
[http://dx.doi.org/10.1080/10408398.2011.599504] [PMID: 24261536]

[28] Isman MB. Plant essential oils for pest and disease management. Crop Prot 2000; (19): 603-8.
[http://dx.doi.org/10.1016/S0261-2194(00)00079-X]

[29] Committee on New Directions in the Study of Antimicrobial Therapeutics. New Classes of Antimicrobials, Committee on New Directions in the Study of Antimicrobial Therapeutics: Immunomodulation, National Research Council (CNDSAT) Treating infectious diseases in a microbial world. National Academies Press 2006.

[30] Fey PD, Safranek TJ, Rupp ME, *et al.* Ceftriaxone-resistant salmonella infection acquired by a child from cattle. N Engl J Med 2000; 342(17): 1242-9.
[http://dx.doi.org/10.1056/NEJM200004273421703] [PMID: 10781620]

[31] Jensen LB, Angulo FJ, Mølbak K, *et al.* Human Health Risks Associated with Antimicrobial Use in Animals. In: Guardabassi L, Jensen LB, Kruse H, Eds. Guide to Antimicrobial Use in Animals. Oxford: UK Blackwell Publishing Ltd 2009.
[http://dx.doi.org/10.1002/9781444302639.ch2]

CHAPTER 2

Bacteriophages as Antibacterial Agents: Why are We Facing an Antibiotic Crisis and How Could Bacteriophages be of Help?

Jean-Paul Pirnay[1], Gilbert Verbeken[1], Mario Vaneechoutte[3], Serge Jennes[2], Maya Merabishvili[1,3,4], Isa Serrano[5], Daniel De Vos[1,*]

[1] *Laboratory for Molecular and Cellular Technology, Queen Astrid Military Hospital, Bruynstraat 1, B-1120, Brussel, Belgium*

[2] *Burn Wound Centre, Queen Astrid Military Hospital, Bruynstraat 1, B-1120, Brussel, Belgium*

[3] *Laboratory for Bacteriology Research, Faculty of Medicine, Ghent University, 9000, Ghent, Belgium*

[4] *George Eliava Institute of Bacteriophages, Microbiology and Virology, 0160, Tbilisi, Georgia*

[5] *Centro de Investigação Interdisciplinar em Sanidade Animal (CIISA), Faculdade de Medicina Veterinária da Universidade de Lisboa, Lisboa, Portugal*

Abstract: Bacteriophages are viral, natural and bacterio-specific entities. As a major part of the biosphere, they were involved in the origin of life. They still play an essential role in evolution and are highly involved in the development of molecular biology. In the issued biotechnology industry, they are promising as a sustainable antibacterial. Felix d'Hérelle, one of the discoverers of bacteriophages, first proposed "(bacterio) phage therapy" in the early 20th century. At the Eliava Institute in Tbilisi, Georgia it was further developed and it is still used in medical practice in all the former Soviet Republics. The Western world, with the advent of antibiotics, almost forgot phage therapy.

The antibiotic resistance crisis placed phage therapy again in the spotlights. The main problem today is the lack of evidence based therapeutic phage studies in accordance to modern standards as well as the lack of an adapted phage therapy regulatory frame. Initiating clinical studies in this context is difficult. Phage therapy is sporadically applied today, although under specific conditions like the Helsinki Declaration and/or specific national regulatory frames (Poland). This impedes clinical application and scientific progress.

[*] **Corresponding author Daniel De Vos:** Laboratory for Molecular and Cellular Technology, Queen Astrid Military Hospital, Bruynstraat 1, B-1120, Brussel, Belgium; E-mail: Daniel.DeVos2@mil.be

Manuela Oliveira and Isa Serrano (Eds)

However, several groups setup animal and human studies, while the idea of using bacteriophages as antibacterial is already applied in the food industry. In the clinic application seems imminent. Several issues, also from the fundamental scientific perspective, still need to be tackled while practically an adapted regulatory frame is urgently needed.

Keywords: Antibiotic crisis, Antibiotic resistance, Bacterial host, Bacteriophage, Diarrhoeal diseases, Eliava Institute, *Escherichia coli*, Food and Drug Administration, Holins, hormesis, *Listeria monocytogenes*, Phage cocktail, Pharmaceutical industry, *Pseudomonas aeruginosa*, Public health, Regulatory obstacles, Safety, Seed lots, *Staphylococcus aureus*, Therapy.

1. INTRODUCTION

Antibiotic resistance (ABR), although not yet fully characterized neither understood, is not a newly discovered biological phenomenon. It is a fact, worldwide, that has a huge negative impact on our societies.

Already Alexandre Fleming, the discoverer of penicillin and initiator of the antibiotic era, noticed the development of ABR. He effectively described ABR and warned the scientific and medical community for this phenomenon in his Nobel Prize lecture in 1945. That was 70 years ago! [1]. ABR originated when bacteria emerged and co-evolved. The discovery and further development of antibiotics brought "western" medicine to the highest level ever. Indeed at the end of the sixties and during the seventies, society and medical officials thought that they could close the book of infectious diseases [2]. However, microbes, such as bacteria, are everywhere and interact with all other living beings. The Spanish microbiologist Baquero and his group described 'antimicrobial resistance' as a 'typical emerging characteristic of a dynamic, highly complex and self-organizing system evolving at the edge of chaos' [3]. Actually, we realize that using antibiotics, once seen as 'miracle drugs', in the way we did and that we are still doing, has resulted in a bacterial response we did not expect. Human interventions in different interconnected fields like human and veterinary medicine, agriculture, the food industry sector, global travel/transport of humans and animals, have accentuated and concentrated the ABR phenomenon worldwide [4]. Unfortunately, we did not do very much to try to minimize or prevent the emergence of enhanced ABR. It is now realized that the ABR phenomenon is not yet completely described and understood. Antibiotics can naturally fulfill different concentration dependent functions (hormesis), among the bacterial communities [5 - 7]. Renowned global and national organizations are just beginning to get grips on the ABR problem [8, 9]. Among the actions and pathways that have recently been proposed to deal with ABR, *one* could be of timely help and this in a sustainable way: phage therapy, or the use of bacterio-specific viruses (bacter - iophages) as antibacterial agents.

In this chapter, we would like to shortly describe the reasons why we are facing the antibiotic crisis and how the bacteriophage approach could be of help in resolving this crisis.

2. WHY ARE WE FACING AN ANTIBIOTIC (AB) CRISIS?

Several causes can explain why we are facing this AB crisis. First, fundamentally, there is still a lack of knowledge of the role of antibiotics in nature and an insufficient integration of the notion of evolution in medicine and in our society as a whole [10, 11]. Second, our actual narrow and too mechanistic view of society and the natural environment, resulted in the over-and misuse of antibiotics in medicine, agriculture, cattle breeding and food production. Why, for example, are we regularly facing very costly epidemics in poultry, cattle or pork meat production? Because farming animals are breed in crowded artificial environments, using clonal species, which does not take into account the fact that variability and diversity is the base of sustainability.

Nature is complex and involves highly dynamic interactions between all its hierarchical structured elements on which evolution interacts.

Yet, man tried to use an antibiotic, a stable static chemical substance, against a living and evolving entity (a bacterial cell). As we know today, and as could have been expected, this resulted in the emergence of bacterial populations that are resistant against that substance. Man continued along the path and used newly discovered or natural or synthetic antibiotics to substitute those antibiotics that had become unusable.

Today, microbiologists are aware that antibiotics have an important role among bacterial populations. Not only do they have a protective (bacterial fighting) role for the producing cell, they also fulfill a concentration dependent (hormesis) signaling role. Antibiotics can induce or inhibit different bacterial responses, influencing bacterial virulence and survival strategies. They even can induce bacterial biofilm formation [12, 13]. Indeed Hoffman *et al.* [12] showed that biofilm formation could simply be an adaptive/surviving strategy of bacteria against antibiotic attacks through the interaction with specific genes like in the case of *Pseudomonas aeruginosa* and aminoglycosides with the aminoglycoside response regulator (*arr*) gene, which results in alterations in the level of c-di-GMP and resistance to the aminoglycoside. Such phenomena were reported in *E. coli* also and not only for subinhibitory antibiotic concentrations but even for some disinfectants like chlorhexidine. A nice review on the subject is the one of Kaplan [13].

It is known that bacteria form biofilms and that a major part of bacterial infections are biofilm based. However, our antibiotics are essentially developed, tested and regulated in the planktonic growth modus. Studies show that there is a significant differential gene expression profile for a bacterium in planktonic modus (free living, unicellular) *versus* the (multicellular) biofilm life-style modus. As a consequence the expected bacterial response to an antibiotic is lowered or even becomes insignificant [14, 15].

The cause of the ABR problem is not limited to the medical field. It is deeply rooted in our society. The majority of antibiotics produced and distributed worldwide are used in the agro-bio and food sector. It is a multifactorial problem touching cultural identities, politics, economy and fundamental biological phenomena [16].

We considered antibiotics once as 'miracle drugs' and neglected their primary diverse role in nature. Antibiotics were developed and marketed without considering future negative outcomes, and this approach worked for a while. But, very quickly, some bacterial populations (in contrast to other manufactured products like cars or cupboards) responded by evolutionary adaptations and some strains evolved into so called mediatized "superbugs". The fact that ABR emerged as a result of 'evolutionary biology' is still not grasped by society.

3. HOW TO RESOLVE THIS ANTIBIOTIC CRISIS?

The pharmaceutical industry has abandoned the development of new antibiotics. The pipeline is nearly dry and this for several reasons, mainly because the development of new drugs is time consuming and costly, while the return on investment is low. Indeed, developing a costly new drug, the use of which will be restricted to limit the emergence of resistance, is commercially unattractive. Without new incentives and conform the 'logic' of our current short sighted pharmaco-economic model, antibiotics seem to have reached their endpoint (Petrella, in the group of Lisbon).

There are many potential solutions to this conundrum, but none of them is likely to be a 'miracle solution'. One possible solution that can be put forward is the development of new antibiotics by public/private partnership (to lower developmental costs).

The acquisition and use of new scientific knowledge (metagenomics, biofilm, quorum sensing studies...), a better implementation of rapid diagnostics (improving the appropriate AB-use) and developing immunologic approaches like immune-modulators and vaccines are other approaches that can be developed in parallel.

All could contribute to help resolve the AB crisis. But all these approaches imply a genuine boost of fundamental research, which will require a change of the research funding systems as well as a change of the actual pharmaco-economic model. This, however, is a worldwide problem, that will take a long time before the relevant societal strata realizes the need to change.

However, there is an approach that looks sustainable, scientifically sense full and timely applicable: 'phage therapy' or the use of bacteriophages as antibacterial agents [17, 18].

4. BACTERIOPHAGES OR IN SHORT 'PHAGES'

Almost a century ago, bacteriophages, ultra-filterable antibacterial entities, were independently discovered in the UK by Twort (1915), and in France by Felix D'Herelle (1917) a French-Canadian microbiologist [19, 20]. It was D'Herelle who came up with the word 'bacteriophage' or 'bacterium eater' and who proposed to use those entities as antibacterial agent. An idea he quickly tried out in practice [20].

Bacteriophages are bacterio-specific viruses that only can directly interact with the prokaryotic bacterial cells [21]. Indeed, the bacterial prokaryotic biochemical machinery enabling the interaction of bacteriophages and bacterial cells is not the same as in the eukaryotic cells that constitute our bodies. For example, the outer membrane structures of bacterial cells, with whom phages first interact in order to be anchored on the bacterial cell wall, as well the polymerases required for the phage genome amplification are not the same in pro- and eukaryotes. This is why bacteriophages cannot directly interact with eukaryotic cells. Bacteriophages are bacterio-specific and as a consequence, in principle, 'safe' for eukaryotic organisms like plants and animals, including humans.

In fact, bacterio-specific bacteriophages, are composed of a nucleic acid genome, double stranded or single stranded, and composed of DNA or RNA, that is packed in a protein capsid. In order to multiply itself the virion, or extracellular viral particle, has to fix itself on a specific bacterium's outer membrane receptor. This receptor interacts with the virion's specific capsid coat or molecular appendages. Once anchored, physico-chemically, the bacteriophage injects its genomic material into the bacterial cell where it will be amplified and capsid packed by the bacterial 'hijacked' specific biochemical machinery. This process results in the production of tens of newly formed bacteriophages that will spread as 'virions' after lysing the infected host bacterium. Indeed the bacteriophage genome encodes for specific enzymes (holins) in order to break open the bacterial cell wall so that the newly produced virions are environmentally dispersed in order to look for infecting again a specific (host) bacterium. As such the cycle can start again while the host bacterium is again killed. This bacterial lysis process of the natural lytic bacteriophages is going to be used in phage therapy. In fact the bacteriophage could be seen as a self-amplifying antibacterial combating specific bacterial pathogens. This lysis process is independent of the antibiotic resistance status of the bacterium. Biofilms, the bacteria's main lifestyle modus, especially in chronic infection, are known not to respond as expected to ABs or to inhibit antibiotic activity by a differential gene expression altering its metabolic status, but not to the bacteriophage activity [15, 22]. As such bacteria and antibiotics could even have a specific synergistic effect against pathogenic bacteria. Indeed, the bacteriophages, by lysing the biofilm matrix structure oblige the bacteria to take on the planktonic lifestyle modus and as such becoming again sensitive for the antibiotic [23]. In this field specifically more research is needed in order to optimize both approaches: bacteriophages to be applied and/or with or without AB's. Also the type of bacteriophages in a cocktail has influence on the outcome of *e.g.* an antibacterial phage cocktail [24 - 26]. Today, after more than 50 years of molecular biology, which was developed for a great part thanks to the study of bacteriophages and another commonly known microbe, the *Escherichia coli* workhorse, we are now in a comfortable position for further developing bacteriophages as antibacterial agents.

They could be used complimentary or as a substitute for antibiotics in the treatment of bacterial infections. Scientific evidence of bacteriophages used as antibacterial agents exists at several levels. There are the theoretical molecular biology insights, beside the *in vitro* laboratory experiments as well as several *in vivo* studies, in animal models and in humans [27 - 30].

5. BACTERIOPHAGE THERAPY OR PHAGE THERAPY

Experience build-up in humans with bacteriophages is almost empiric and originating from the former Soviet Union Republics where the Eliava Institute in Tbilisi (Georgia) became the main centre of phage therapy. Also in Russia phage therapy is actually not forgotten as well described by Reardon [31]. In the EU, in Poland, phage therapy is applied under specific legislative frames. The Polish Ludwik Hirszfeld Institute of Immunology and Experimental Therapy, based in Wroclaw, is emerging as a very active research unit (clinical phage therapy center included). Their research efforts and clinical phage therapy experience will certainly help to re-introduce phage therapy in accordance with "western" standards of biomedical research and care [28, 29]. In other countries several human clinical trials, compliant with modern biomedical standards, were launched while well described clinical case reports are sporadically published [26, 29, 32 - 36].

6. EXPERIMENTAL PHAGE THERAPY

A well-defined and quality controlled bacteriophage cocktail, BFC-1, active against the gram-negative bacterium *P. aeruginosa* and the gram-positive *Staphylococcus aureus* strains that populated the burn wound centre of the Queen Astrid Military Hospital, was developed in 2007 [35]. Recently, this phage cocktail was tested in nine patients with acute burn wounds colonized with multi-drug resistant (MDR) *P. aeruginosa* and/or *S. aureus*, after the Ethical Committee approval [26]. In this Hospital, phage therapy is considered as part of a save therapy for patients in the condition of therapeutic dead end, particularly for those with infections caused by multi-resistant microorganisms.

Although there was not an adequate evaluation of the efficacy of the phage cocktail, mainly due to the small sample, it revealed several drawbacks:

- The biopsy sampling hampered the clinical trial. In future clinical trials, burn wound colonization/infection monitoring should be done by a semi-quantitative swab cultures procedure in conjunction with clinical signs instead of the textbook gold standard of quantitative biopsy cultures since the latter is too high in workload, while in fact it gives only a snapshot of a small wound localization.
- There was an unsatisfactory bacterial load of wound tissue at the moment of BFC-1 application. Clinical signs and biological markers of burn wound colonization would be more useful in the future than delayed bacteriological results and should be included as a criterion for this type of trial.
- Finally, the sprayed BFC-1 phage cocktail had the tendency to overflow the burn wound. The use of a suitable carrier compatible with phage activity, such as a gel or a dressing, would be more appropriate.

This first experimental phage cocktail therapy was particularly relevant in three aspects: first, it helped to overcome the psychological hurdle to the use of viruses in patients therapy; second, some drawbacks were detected which will help to take measures for future experimental phage therapy; and third, it was determined that the local topical application of BFC-1 in the nine patients did not elicit any adverse event, clinical abnormalities, or changes in laboratory test results [26].

P. aeruginosa is the most prevalent pathogen in cystic fibrosis (CF) patients and increasingly resistant to antibiotics [37], turning difficult its treatment with classical antibiotherapy. Most of these patients will eventually evolve in to chronic pulmonary bacterial infections and die [38, 39]. In a recent experimental phage therapy, the efficacy of a cocktail composed by ten bacteriophages against *P. aeruginosa* following its addition to 58 sputum samples from patients with CF collected at three French hospitals were assessed [40]. The bacteriophages were not specifically selected for their ability to infect *P. aeruginosa* strains in sputum.

P. aeruginosa was detected in most samples (n = 48). In these samples, the addition of the cocktail led to a significant decrease in the number of bacteria. The number of bacteriophages increased over the threshold value in 86.4% of samples. Thus, the microenvironment observed in the lungs of CF patients does not prevent the activity of bacteriophages.

Each of the ten bacteriophages were individually tested against 20 bacterial colonies and in 64.6% of the samples it was detected a bacteria susceptible to a phage. This confirms that, due to clonal mutations, different phenotypes can emerge from a unique bacterial genotype in the sputum of one patient. Therefore, testing a set of bacteriophages against a group of colonies for each patient would be the best approach for formulating bacteriophage cocktails.

The authors clearly demonstrated that the most efficient bacteriophages *in vivo* are most likely those isolated using the patients' strains. This suggests the use of a personalized therapy [41, 42]. The authors also demonstrated that bacteriophages infect bacteria regardless of the clinical properties of the patients. This conclusion goes towards the development of a bacteriophage therapy to treat chronic lung infections [40].

7. SAFETY ISSUES

In fact a bacteriophage is a self-amplifying antibacterial which will be naturally eliminated from the body by our reticulo-endothelial system, the liver Kuppfer cells, the urine and the faeces, once the targeted bacterial host cells disappears from the infectious site.

Are bacteriophages, as viruses, safe for therapeutic treatment of patients (animals or humans)? This is one of the first questions people ask when they hear that viruses will be put on patients for destroying pathogenic bacteria.

8. IN FACT WE LIVE IN AN "OCEAN" OF BACTERIOPHAGES

Research shows that bacteriophages are the most abundant biological entities in our planetary biosphere [43 - 46]. Where bacteria are present, the bacteriophages also are present outnumbering them in at least a factor 10 to 100. All our life we eat and drink bacteriophages while our body carries bacteriophages. Bacteria themselves, present in our body, already exceed the number of human cells constituting this body.

Bacteriophages can be seen as bacterial predators in a continuous co-evolutionary arms race. As such they regulate the earth's ecological equilibrium in several environmental or ecological niches. We should start thinking from the co-evolving couplet phage/bacterium, continually co-adapting to each other. This is how and why phage therapy provides a long-term sustainable antibacterial treatment approach.

For each existing or emerging pathogen there exists or will emerge a bacteriophage, making bacteriophage therapy sustainable. This ecological perspective endowing also the notion of evolution is well fitting in the emerging field of evolutionary/Darwinian medicine. It is also an approach completely in line with the societal and political interest(s) in long-term sustainable approaches required for managing our globalizing world [47] where numbers of diseases and epidemics will increase.

Bacteriophages exist since the bacterial cells originated which makes them the other facet of a bacterium. Raoult and Forterre brought viruses, including bacteriophages, into the tree of life. We could consider viruses as living biological entities, although from another kind as we are used [48].

Studies show that half of the bacterial biomass daily is lysed by lytic bacteriophages, whose total number in our biosphere is estimated to approach 10^{31}.

The group of Rohwer showed that all epithelial mucus layers in metazoan, such as our own gut mucus layer, is in fact a bacteriophage based symbiotic defense system that protects us against potential invading bacteria [49]. This study shows us the inherent and basic inoffensiveness of bacteriophages and argues for safety of bacteriophages used as antibacterial.

In the seventies the Food and Drug Administration (FDA) setup a review of safety of bacteriophages for humans because bacteriophages were found into several vaccine preparations. The conclusion of this study review was that bacteriophages were not a safety issue. This allowed the use in humans of the concerned vaccines containing bacteriophages [50].

The application of phage phiX174 intravenously in HIV positive patients later on was also allowed. The FDA concluded that bacteriophages are safe for use in humans.

Taking this into consideration, in conjunction with the existing empirical clinical experience as well as several studies in diverse animals and a handful recent human applications, demonstrate that the phage therapy concept is intrinsically safe [50 - 52].

9. SAFETY REQUIREMENTS FOR PHAGE THERAPY

Experts agree that clinical regulated trials are required [58] although that those clinical trial regulation should be adapted in order to make phage therapy a real sustainable approach applicable also for individual or specific patients in need. According to phage experts [53], phage therapy products should not be exclusively developed and marketed as antibiotics and applying the economic values of the pharmaceutics industry. These would allow a timely and less expensive supplying of natural bacteriophages for a personalized therapy. In the European Union (EU), human tissues and cells that are not considered as advanced therapy medicinal products (ATMPs) are exclusively regulated by the EU tissue and cell directives (EUTCDs). Phage therapy products may be included in this category. Consequently, a set of requirements to be applied to the production of phage therapy products were defined by thirty two phage experts from 12 different countries, and based on the EUTCDs quality and safety standards for human cells. The document enumerates all possible quality and safety risks related to phage therapy products, as well as the tests that can be performed to minimize those risks. However, they do not attend the efficacy aspects of phage products.

The quality and safety requirements for sustainable phage therapy products include:

1. Production environment. The specified air quality and sanitation must be validated and monitored in order to minimize the risk of contamination;
2. Production processes equipment and materials. In summary, all equipment and material must be designed and maintained according to their purpose and must minimize any hazard. They must be identified and validated, inspected and preventively maintained in agreement with the manufacturers' instructions. Standard operating procedures (SOPs) must detail the specifications for production processes, all critical materials and reagents and must be validated. Bacteria and phage bank systems (master seed lots and working seed lots) need to be set up;
3. Quality assurance and quality control specifications:
    a. Host bacteria used in production (stock suspensions). The bacterial hosts used in the production process should be as safe or least pathogenic as possible (Table **1**).

Table 1. Characteristics, control tests, and recommended tests of bacterial stock suspensions [53].

Characteristics	**Control test**	**Limits of acceptance**	**Recommended test**
Origin	• Document pedigree, history, pathogenicity level	• Known origin	• Screening of scientific literature
Identification	• Identification at the species and strain levels	• Matching species and strain identification	• Clinical microbiology techniques • Highly discriminating typing techniques
Try to use non-lysogenic strains, containing as few phages (or other phage-like elements) as possible	• Induction of phages	• As few spontaneously produced (or by induction) as possible	• *In vitro* induction methods (Mitomycine C [54] or UV induction)
	• Host genome screening		• State of the art DNA sequencing and analysis (bioinformatics) procedures

(Table 1) contd.....

Characteristics	Control test	Limits of acceptance	Recommended test
Avoid mutator strains as host bacteria	• Screen for mutator strains	• No mutator strain	• State of the art tests (*e.g.* fosfomycin and rifampicin disk diffusion tests) [55]
Validated preservation, storage (*e.g.* freeze-drying)	• Monitor storage conditions (*e.g.* temperature)	• Variable, depending on the preservation method	• Variable (*e.g.* temperature probes)

b. Bacteriophages (master seed lots). One requirement for master seed lots is the absence of potentially damaging genetic determinants such as those conferring toxicity, virulence, lysogeny or antibiotic resistance (Table **2**).

Table 2. Characteristics, control tests, and recommended tests of master seed lots [53].

Characteristics	Control test	Limits of acceptance	Recommended test
Origin		• Known origin	
	• Document bacteriophage pedigree, history (*e.g.* isolation source)		• Screening of scientific literature
		• Natural or naturally evolved bacteriophages	
Identification	• Identification at the family, genus, species and strain levels	• Matching identification, morphology and biology	• State of the art DNA or RNA sequencing and analysis procedures
	• Morphology and biology		• Highly discriminating genotyping techniques • State of the art classification according to the International Committee on Taxonomy of Viruses (ICTV) • One step growth curve [56]
Not containing potentially damaging genetic determinants	• Genome analysis for known potentially damaging genetic determinants	• Absence of potentially damaging genetic determinants	• State of the art DNA or RNA sequencing and genome analysis (bioinformatics) procedures

(Table 2) contd.....

Characteristics	Control test	Limits of acceptance	Recommended test
In vitro efficacy	• Determination of host range on a panel of target reference strains	• Broad host range (if possible)	• Titration of bacteriophages against target bacteria according to the soft-agar overlay method [57]
		• Variable threshold according to species	• Spot test [56]
	• Determination of frequency of emergence of phage resistant bacteria	• Low frequency of emergence of resistance	• Method described by Adams [57]
Improvement, adaptation, "training" (if necessary)	• Optimization of host range	• Broadened and stable host range	• Titration of bacteriophages against target bacteria according to the soft-agar overlay method [56] • Spot test [56]
Validated preservation, storage (*e.g.* freeze-drying)	• Monitor storage conditions (*e.g.* temperature)	• Variable, depending on the preservation method	• Variable (*e.g.* temperature probes)

c. Bacteriophages (working seed lots and active substances) (Table **3**).

Table 3. Characteristics, control tests, and recommended tests of working seed lots and active substances [53].

Characteristics	Control test	Limits of acceptance	Recommended test
Quantitative determination of active substance	• Bacteriophage titration	• Variable. Typically log(8) to log(10) plaque forming units (pfu)/ml	• Soft-agar overlay method [57]
Identification of active substance	• Genomic fingerprinting	• Matching genomic fingerprint	• State of the art genotyping techniques
Microbial contamination	• Sterility (when there is no urgency)	• Sterile (absence of micro-organisms)	• Membrane filtration method based on the European Pharmacopoeia (EP)
	• Absence of pathogens (when urgent)	• Aseptic (absence of pathogens)	• State of the art clinical microbiology methods
Toxicity	• Bacterial endotoxin or lipopolysaccharides quantification [57]	• Depends on posology and method and route of administration.	• Limulus Amebocyte Lysate (LAL) assay according to the EP (*e.g.* kinetic-QCL method)

(Table 3) contd.....

Characteristics	Control test	Limits of acceptance	Recommended test
Bacterial DNA contamination	• Screen for (potentially damaging) host bacterial DNA	• Absence of potentially damaging genetic determinants that are known to be present in the host bacterium	• Methods for the quantification of bacterial DNA in general
Acidity or basicity of aqueous solution	• pH measurement	• Variable (typically 6,5 to 7,5)	• pH test (EP method)
Purity	• Clarity of phage solution	• Absence of visible particles	• EP method, CPMP-ICH guideline
Validated preservation, storage (*e.g.* freeze-drying)	• Monitor, record, demonstrate storage conditions	• Variable (*e.g.* 2–8°C)	• Variable (*e.g.* temperature probes)

   d. Finished products;

4. Shelf life of phage stock suspensions, working solutions and finished products (at recommended storage conditions).
5. Surveillance. The clinical use of phage therapy products must be surveyed and reported, together with possible undesirable reactions associated with the use of phage therapy. A centralized and public open reporting system is necessary [53].

10. THE POTENTIAL ADVANTAGES OF PHAGE THERAPY *VERSUS* ANTIBIOTICS

Seen the bacterio specificity of bacteriophages and the accumulating evidence through studies, it can be stated that the collateral damage on different microbiota eventually created by the use of phages is at least lower or non-existent in comparison with the use of antibiotics. Even small spectrum antibiotics affect negatively the microbiota. The (side) effects of using AB on our natural microbiota are so huge that it seems that a lot of actual diseases are associated with a microbiome disturbance caused by the use of AB. This aspect is well documented and described globally by Blaser [59].

The use of bacteriophages could prevent antibiotic side effects like *Clostridium difficile* infections or other gastrointestinal diseases linked to a dysbiosis or unbalanced gut microbiota. Bacteriophages could eventually restore or re-equilibrate a disturbed situation.

Another advantage of the use of bacteriophages as an antibacterial is the fact that looking for a specific phage against a specific bacterium is possible in a much shorter timeframe (days to weeks) than searching and developing a new AB for treating the same problem [60 - 62]. The phage-treatment should, at least theoretically, also be cheaper. This specifically makes phage therapy a relatively cheap 'online' approach in the fight against bacterial infections.

11. PHAGE THERAPY IN PUBLIC HEALTH

Individual hospitalized patients are not the only patients who are affected by ABR. All types of healthcare facilities (like *e.g.* homes for elderly) and even whole communities can be confronted with serious - eventually untreatable health problems due to ABR issues.

We bring in remembrance the Entero-aggregative Hemorrhagic *Escherichia coli* (EAHEC) 0104:H4 outbreak in Berlin in 2011. The value of antibiotics for treating such infections is a questionable issue [60].

Taking the Berlin 2011 epidemic as an example it was shown, by different groups, that potent lytic bacteriophages virulent for this problematic entero-aggregative *E. coli*, could be relatively quickly obtained either by isolation of new phages from the environment or by selection of phages from existing collections. Candidate therapeutic bacteriophages were obtained in a matter of days. This fact illustrates that bacteriophages could have their value helping to control this O104:H4 *E. coli* outbreak. Unfortunately, the authorized use of bacteriophages was not possible in this context. Indeed under the existing medicinal product legislation such an anti-O104:H4 phage preparation would have taken years to develop, produce and to get approved for therapeutic use on patients. Seen the potential, also for public health purposes of bacteriophages, it seems of uttermost importance to set up therapeutic phage collections and to maintain and continuously update these bacteriophage collection(s), which can be used to control bacterial epidemics in a timely and theoretical cost-effectively way.

Bacterial diarrhoeal diseases are, in large parts of the world, a worrisome big issue in public health. Many common causative agents are MDR bacteria which can especially in poor sanitation environments, quickly provoke epidemics and pandemics. Also pandemics of cholera can be referred in this context. Bacteriophages have the potential to play a significant role in tackling this major health issue [63].

The group of Mekalanos *et al*. showed the effective natural role of bacteriophages as the agents at the origin of the natural collapse of local endemic diarrhoeal epidemics in Bangladesh [64, 65]. This observation brings us to the statement that further study of the bacterial/phage interaction could bring phage therapy, as a combat-tool to ABR, to the attention of the public health authorities worldwide. It would be a supplementary way to control or at least inhibit diarrhoeal diseases worldwide and also in developing countries.

12. WHY PHAGE THERAPY IS NOT YET IMPLEMENTED?

Bacteriophages, being natural biological entities that co-evolve with their bacterial host cells, are definitely not of the same kind as classic drugs. They are very different than a chemical stable molecule, like antibiotics. Phage therapy obliges us to think of bacteriophages and bacteria as a couplet in continuous interaction. Our actual ("western") pharmaco-economic model, that is mainly based on development and market-placement of stable static chemical molecules, evolved into a highly sophisticated regulatory network that hampers the (re)introduction of phage therapy into the actual "western" medicinal world. Phage therapy, in contrast to the use of classic antibiotics, is based on auto-amplification and co-evolution of a viral natural entity and its bacterial host cell [61, 62, 66, 67].

Beside this the intellectual property rights, an essential aspect for industry, are a thorny issue. Patents could be granted and exist, but how robust will they be?

Altogether, because of the serious negative societal impact of ABR, the phage therapy concept should be taken seriously and as a realistic tool to resolve the actual worldwide antibiotic crisis.

Currently the Small and Medium Enterprises (SME's) dealing with bacteriophages face several hurdles in order to put bacteriophage products as therapeutic agents on the market, especially in the EU and US.

Some companies therefore decided to circumvent some regulatory obstacles by first going into the agricultural and food applications. In this area regulations are less stringent. As such since, in 2006 the FDA approved a bacteriophage preparation on ready-to-eat meat and poultry products as an antimicrobial against *Listeria monocytogenes*.

CONCLUSIONS AND PERSPECTIVES

Conclusively we can state that the use of bacteriophages as antibacterial agent makes sense. Phage therapy needs to be seen as an additional complimentary eventually synergistic tool to resolve the actual ABR crisis. Phage therapy will not only be relevant to help individual hospitalized patients or patients present in other healthcare settings. Phage therapy could also be a beneficial public health tool for inhibiting and/or to control emerging bacterial epidemics, especially in developing countries where bacterial diarrhoeal disease, amongst others, takes a major part of the infectious burden.

Phage therapy can be used in conjunction or without AB's, in relation to the specific indication. In general, a more judicious use of antibiotics should be promoted.

All future antibacterial strategies should be integrated, which means the involvement of different fields of concern such as human and veterinary medicine, the agro-bio and food industry, the environment in general.

Our actual worldview, and our actual pharmaco-economic model that goes with this view, should be adapted. Only taking all this into account as an antibacterial treatment option, phage therapy could be further developed, optimized and implemented.

CONFLICT OF INTEREST

The authors confirm that this chapter contents have no conflict of interest.

ACKNOWLEDGEMENTS

The authors wish to thank the support of the Belgian Ministry of Defense, especially General Geert Laire and General Pierre Neirinckx .

REFERENCES

[1] Fleming A, Ed. Penicillin. Nobel prize Lecture. http://www.nobelprize.org/nobel_prizes/medicine/laureates/1945/fleming-lecture.pdf. 1945.

[2] Allen HK, Donato J, Wang HH, Cloud-Hansen KA, Davies J, Handelsman J. Call of the wild: antibiotic resistance genes in natural environments. Nat Rev Microbiol 2010; 8(4): 251-9. [http://dx.doi.org/10.1038/nrmicro2312] [PMID: 20190823]

[3] Levy SB, Marshall B. Antibacterial resistance worldwide: causes, challenges and responses. Nat Med 2004; 10(12) (Suppl.): S122-9. [http://dx.doi.org/10.1038/nm1145] [PMID: 15577930]

[4] Rice LB. Federal funding for the study of antimicrobial resistance in nosocomial pathogens: no ESKAPE. J Infect Dis 2008; 197(8): 1079-81. [http://dx.doi.org/10.1086/533452] [PMID: 18419525]

[5] Theophel K, Schacht VJ, Schlüter M, *et al.* The importance of growth kinetic analysis in determining bacterial susceptibility against antibiotics and silver nanoparticles. Front Microbiol 2014; 5: 544. [http://dx.doi.org/10.3389/fmicb.2014.00544] [PMID: 25426104]

[6] Davies J, Spiegelman GB, Yim G. The world of subinhibitory antibiotic concentrations. Curr Opin Microbiol 2006; 9(5): 445-53. [http://dx.doi.org/10.1016/j.mib.2006.08.006] [PMID: 16942902]

[7] Calabrese EJ, Hoffmann GR, Stanek EJ, Nascarella MA. Hormesis in high-throughput screening of antibacterial compounds in *E. coli*. Hum Exp Toxicol 2010; 29(8): 667-77. [http://dx.doi.org/10.1177/0960327109358917] [PMID: 20068011]

[8] ECDC, EMEA. The bacterial challenge: time to react. A call to narrow the gap between multidrug-resistant bacteria in the EU and the development of new antibacterial agents Stockholm 2009.

[9] WHO. Antimicrobial resistance: Global report on Surveillance. Geneva. 2014.

[10] Baquero F, Coque TM, Canton R. Antibiotics, complexity and evolution. ASM News 2003; 69: 547-52.

[11] Bassetti M, Ginocchio F, Mikulska M, Taramasso L, Giacobbe DR. Will new antimicrobials overcome resistance among Gram-negatives? Expert Rev Anti Infect Ther 2011; 9(10): 909-22. [http://dx.doi.org/10.1586/eri.11.107] [PMID: 21973303]

[12] Hoffman LR, D'Argenio DA, MacCoss MJ, Zhang Z, Jones RA, Miller SI. Aminoglycoside antibiotics induce bacterial biofilm formation. Nature 2005; 436(7054): 1171-5. [http://dx.doi.org/10.1038/nature03912] [PMID: 16121184]

[13] Kaplan JB. Antibiotic-induced biofilm formation. Int J Artif Organs 2011; 34(9): 737-51. [http://dx.doi.org/10.5301/ijao.5000027] [PMID: 22094552]

[14] Beloin C, Renard S, Ghigo JM, Lebeaux D. Novel approaches to combat bacterial biofilms. Curr Opin Pharmacol 2014; 18: 61-8. [http://dx.doi.org/10.1016/j.coph.2014.09.005] [PMID: 25254624]

[15] Beloin C, Valle J, Latour-Lambert P, *et al.* Global impact of mature biofilm lifestyle on *Escherichia coli* K-12 gene expression. Mol Microbiol 2004; 51(3): 659-74.

[http://dx.doi.org/10.1046/j.1365-2958.2003.03865.x] [PMID: 14731270]

[16] Pirnay JP, De Vos D, Zizi M, Heyman P. No easy way to exterminate 'superbugs' at the dawn of the third millennium. Expert Rev Anti Infect Ther 2003; 1(4): 523-5. [http://dx.doi.org/10.1586/14787210.1.4.523] [PMID: 15482147]

[17] Bush K, Courvalin P, Dantas G, *et al.* Tackling antibiotic resistance. Nat Rev Microbiol 2011; 9(12): 894-6. [http://dx.doi.org/10.1038/nrmicro2693] [PMID: 22048738]

[18] Thiel K. Old dogma, new tricks-21st Century phage therapy. Nat Biotechnol 2004; 22(1): 31-6. [http://dx.doi.org/10.1038/nbt0104-31] [PMID: 14704699]

[19] Twort FW. investigation on the nature of the ultramicroscopic viruses. Lancet 1915; 186: 1241-3. [http://dx.doi.org/10.1016/S0140-6736(01)20383-3]

[20] d'Herelle F. Sur le rôle du microbe bactériophage dans la typhose aviaire. C R Acad Sci 1919; 169: 932-4.

[21] Sulakvelidze A, Kutter E. Bacteriophage therapy in humans. In: Kutter E, Sulakvelidze A, Eds. Bacteriophages: Biology and Application. Boca Raton, Florida: CRC Press 2005; pp. 381-436.

[22] Whiteley M, Bangera MG, Bumgarner RE, *et al.* Gene expression in *Pseudomonas aeruginosa* biofilms. Nature 2001; 413(6858): 860-4. [http://dx.doi.org/10.1038/35101627] [PMID: 11677611]

[23] Comeau AM, Tétart F, Trojet SN, Prère MF, Krisch HM. Phage-Antibiotic Synergy (PAS): beta-lactam and quinolone antibiotics stimulate virulent phage growth. PLoS One 2007; 2(8): e799. [http://dx.doi.org/10.1371/journal.pone.0000799] [PMID: 17726529]

[24] Bourdin G, Navarro A, Sarker SA, *et al.* Coverage of diarrhoea-associated *Escherichia coli* isolates from different origins with two types of phage cocktails. Microb Biotechnol 2014; 7(2): 165-76. [http://dx.doi.org/10.1111/1751-7915.12113] [PMID: 24528873]

[25] Essoh C, Blouin Y, Loukou G, *et al.* The susceptibility of *Pseudomonas aeruginosa* strains from cystic fibrosis patients to bacteriophages. PLoS One 2013; 8(4): e60575. [http://dx.doi.org/10.1371/journal.pone.0060575] [PMID: 23637754]

[26] Rose T, Verbeken G, Vos DD, *et al.* Experimental phage therapy of burn wound infection: difficult first steps. Int J Burns Trauma 2014; 4(2): 66-73. [PMID: 25356373]

[27] Brüssow H. Phage therapy: the *Escherichia coli* experience. Microbiology 2005; 151(Pt 7): 2133-40. [http://dx.doi.org/10.1099/mic.0.27849-0] [PMID: 16000704]

[28] Chanishvili N, Sharp R. A literature review of the practical application of bacteriophage research Eliava Institute of Bacteriophage, Microbiology and Virology. Tbilisii, Georgia: Eliava Foundation 2009.

[29] Górski A, Miedzybrodzki R, Borysowski J, *et al.* Bacteriophage therapy for the treatment of infections. Curr Opin Investig Drugs 2009; 10(8): 766-74. [PMID: 19649921]

[30] Smith HW, Huggins MB. Successful treatment of experimental *Escherichia coli* infections in mice

using phage: its general superiority over antibiotics. J Gen Microbiol 1982; 128(2): 307-18. [PMID: 7042903]

[31] Reardon S. Phage therapy gets revitalized. Nature 2014; 510(7503): 15-6. [http://dx.doi.org/10.1038/510015a] [PMID: 24899282]

[32] Therapy P. Page Therapy: Current Research and Applications. Poland: Caister Academic press 2014.

[33] Khawaldeh A, Morales S, Dillon B, *et al.* Bacteriophage therapy for refractory *Pseudomonas aeruginosa* urinary tract infection. J Med Microbiol 2011; 60(Pt 11): 1697-700. [http://dx.doi.org/10.1099/jmm.0.029744-0] [PMID: 21737541]

[34] Kutter E, De Vos D, Gvasalia G, *et al.* Phage therapy in clinical practice: treatment of human infections. Curr Pharm Biotechnol 2010; 11(1): 69-86. [http://dx.doi.org/10.2174/138920110790725401] [PMID: 20214609]

[35] Merabishvili M, Pirnay JP, Verbeken G, *et al.* Quality-controlled small-scale production of a well-defined bacteriophage cocktail for use in human clinical trials. PLoS One 2009; 4(3): e4944. [http://dx.doi.org/10.1371/journal.pone.0004944] [PMID: 19300511]

[36] Rhoads DD, Wolcott RD, Kuskowski MA, *et al.* Bacteriophage therapy of venous leg ulcers in humans: results of a phase I safety trial. J Wound Care 2009 Jun; 18(6): 237-8-40-3. [PMID: 19661847]

[37] Croughs PD, Li B, Hoogkamp-Korstanje JA, Stobberingh E. Thirteen years of antibiotic susceptibility surveillance of *Pseudomonas aeruginosa* from intensive care units and urology services in the Netherlands. Eur J Clin Microbiol Infect Dis 2013; 32(2): 283-8. [http://dx.doi.org/10.1007/s10096-012-1741-4] [PMID: 22960982]

[38] Hansen CR, Pressler T, Høiby N. Early aggressive eradication therapy for intermittent *Pseudomonas aeruginosa* airway colonization in cystic fibrosis patients: 15 years experience. J Cyst Fibros 2008; 7(6): 523-30. [http://dx.doi.org/10.1016/j.jcf.2008.06.009] [PMID: 18693078]

[39] Høiby N. Recent advances in the treatment of *Pseudomonas aeruginosa* infections in cystic fibrosis. BMC Med 2011; 9: 32. [http://dx.doi.org/10.1186/1741-7015-9-32] [PMID: 21463524]

[40] Saussereau E, Vachier I, Chiron R, *et al.* Effectiveness of bacteriophages in the sputum of cystic fibrosis patients. Clin Microbiol Infect 2014; 20(12): O983-90. [http://dx.doi.org/10.1111/1469-0691.12712] [PMID: 24920209]

[41] Chan BK, Abedon ST, Loc-Carrillo C. Phage cocktails and the future of phage therapy. Future Microbiol 2013; 8(6): 769-83. [http://dx.doi.org/10.2217/fmb.13.47] [PMID: 23701332]

[42] Henry M, Lavigne R, Debarbieux L. Predicting *in vivo* efficacy of therapeutic bacteriophages used to treat pulmonary infections. Antimicrob Agents Chemother 2013; 57(12): 5961-8. [http://dx.doi.org/10.1128/AAC.01596-13] [PMID: 24041900]

[43] Bamford DH. Do viruses form lineages across different domains of life? Res Microbiol 2003; 154(4): 231-6. [http://dx.doi.org/10.1016/S0923-2508(03)00065-2] [PMID: 12798226]

[44] Bergh O, Børsheim KY, Bratbak G, Heldal M. High abundance of viruses found in aquatic environments. Nature 1989; 340(6233): 467-8.
[http://dx.doi.org/10.1038/340467a0] [PMID: 2755508]

[45] Fuhrman JA. Marine viruses and their biogeochemical and ecological effects. Nature 1999; 399(6736): 541-8.
[http://dx.doi.org/10.1038/21119] [PMID: 10376593]

[46] Hendrix RW. Bacteriophages: evolution of the majority. Theor Popul Biol 2002; 61(4): 471-80.
[http://dx.doi.org/10.1006/tpbi.2002.1590] [PMID: 12167366]

[47] Shanks N, Pyles RA. Evolution and medicine: the long reach of "Dr. Darwin". Philos Ethics Humanit Med 2007; 2: 4.
[http://dx.doi.org/10.1186/1747-5341-2-4] [PMID: 17407602]

[48] Raoult D, Forterre P. Redefining viruses: lessons from Mimivirus. Nat Rev Microbiol 2008; 6(4): 315-9.
[http://dx.doi.org/10.1038/nrmicro1858] [PMID: 18311164]

[49] Barr JJ, Auro R, Furlan M, *et al.* Bacteriophage adhering to mucus provide a non-host-derived immunity. Proc Natl Acad Sci USA 2013; 110(26): 10771-6.
[http://dx.doi.org/10.1073/pnas.1305923110] [PMID: 23690590]

[50] Sulakvelidze A. The challenges of bacteriophage therapy. Eur Ind Pharm 2011; 10: 14-8.

[51] Biswas B, Adhya S, Washart P, *et al.* Bacteriophage therapy rescues mice bacteremic from a clinical isolate of vancomycin-resistant Enterococcus faecium. Infect Immun 2002; 70(1): 204-10.
[http://dx.doi.org/10.1128/IAI.70.1.204-210.2002] [PMID: 11748184]

[52] Bruttin A, Brüssow H. Human volunteers receiving *Escherichia coli* phage T4 orally: a safety test of phage therapy. Antimicrob Agents Chemother 2005; 49(7): 2874-8.
[http://dx.doi.org/10.1128/AAC.49.7.2874-2878.2005] [PMID: 15980363]

[53] Pirnay JP, Blasdel BG, Bretaudeau L, *et al.* Quality and safety requirements for sustainable phage therapy products. Pharm Res 2015; 32(7): 2173-9.
[http://dx.doi.org/10.1007/s11095-014-1617-7] [PMID: 25585954]

[54] Burlage RS, Atlas R, Stahl D, *et al.* Techniques in microbial ecology. USA: Oxford University Press 1998.

[55] Galán JC, Tato M, Baquero MR, Turrientes C, Baquero F, Martinez JL. Fosfomycin and rifampin disk diffusion tests for detection of *Escherichia coli* mutator strains. J Clin Microbiol 2004; 42(9): 4310-2.
[http://dx.doi.org/10.1128/JCM.42.9.4310-4312.2004] [PMID: 15365030]

[56] Kutter E, Sulakvelidze A. Bacteriophages: Biology and application. Boca Raton: CRC Press 2005.

[57] Adams MH. Bacteriophages. New York: Interscience Publishers 1959.

[58] Parracho HM, Burrowes BH, Enright MC, McConville ML, Harper DR. The role of regulated clinical trials in the development of bacteriophage therapeutics. J Mol Genet Med 2012; 6: 279-86.
[http://dx.doi.org/10.4172/1747-0862.1000050] [PMID: 22872803]

[59] Blaser M. Missing Microbes: How the overuse of antibiotics is fuelling our modern Plagues. New York: Henry Holt and Company 2014.

[60] Merabishvili M, De Vos D, Verbeken G, *et al.* Selection and characterization of a candidate therapeutic bacteriophage that lyses the *Escherichia coli* O104:H4 strain from the 2011 outbreak in Germany. PLoS One 2012; 7(12): e52709.
[http://dx.doi.org/10.1371/journal.pone.0052709] [PMID: 23285164]

[61] Pirnay JP, De Vos D, Verbeken G, *et al.* The phage therapy paradigm: prêt-à-porter or sur-mesure? Pharm Res 2011; 28(4): 934-7.
[http://dx.doi.org/10.1007/s11095-010-0313-5] [PMID: 21063753]

[62] Pirnay JP, Verbeken G, Rose T, *et al.* Introducing yesterday's phage therapy in today's medicine. Future Virol 2012; 7: 379-90.
[http://dx.doi.org/10.2217/fvl.12.24]

[63] Antibiotic resistance in developing countries. New York: Springer 2010.

[64] Faruque SM, Islam MJ, Ahmad QS, *et al.* Self-limiting nature of seasonal cholera epidemics: Role of host-mediated amplification of phage. Proc Natl Acad Sci USA 2005; 102(17): 6119-24.
[http://dx.doi.org/10.1073/pnas.0502069102] [PMID: 15829587]

[65] Faruque SM, Naser IB, Islam MJ, *et al.* Seasonal epidemics of cholera inversely correlate with the prevalence of environmental cholera phages. Proc Natl Acad Sci USA 2005; 102(5): 1702-7.
[http://dx.doi.org/10.1073/pnas.0408992102] [PMID: 15653771]

[66] Verbeken G, De Vos D, Vaneechoutte M, Merabishvili M, Zizi M, Pirnay JP. European regulatory conundrum of phage therapy. Future Microbiol 2007; 2(5): 485-91.
[http://dx.doi.org/10.2217/17460913.2.5.485] [PMID: 17927471]

[67] Verbeken G, Pirnay JP, De Vos D, *et al.* Optimizing the European regulatory framework for sustainable bacteriophage therapy in human medicine. Arch Immunol Ther Exp (Warsz) 2012; 60(3): 161-72.
[http://dx.doi.org/10.1007/s00005-012-0175-0] [PMID: 22527355]

CHAPTER 3

Antimicrobial Peptides

Isa Domingues Serrano*

Centro de Investigação Interdisciplinar em Sanidade Animal (CIISA), Faculdade de Medicina Veterinária da Universidade de Lisboa, Avenida da Universidade Técnica, 1300-477 Lisboa, Portugal

Abstract: Most Antimicrobial Peptides (AMPs) are small peptides with 10–40 residues. They are part of the immune system in virtually all multicellular organisms as a defense strategy. They are promising drug candidates for the control or treatment of several diseases or to be used in combination with conventional therapy. There are several AMPs effective against cancer cells and diseases caused by bacteria, fungi and parasites. However, none AMP is currently commercialized due to a number of reasons such as high production costs as compared to conventional antibiotics. The new generation of AMPs, mainly antimicrobial peptidomimetics, has several drug candidates currently in Phase I and II clinical trials. Understanding the AMPs' fully mechanism of action is crucial to develop AMPs into useful antimicrobial agents highlighting their strengths: broad-spectrum antimicrobial activity, low probability to generate resistance, and potential for topical and injected applications.

Keywords: Amphipathicity, Anticancer, Antimicrobial peptides, Antiparasitic, Antiviral, Cathelicidin, Cationicity, Conventional antibiotics, Databases, Defensins, Dissemination, Drug development, Immune system, Interaction, Microorganisms, Multiresistance, Peptidomimetics, Permeabilization, Pharmaceutical companies, Therapy.

INTRODUCTION

The prevalence of drug-resistant microorganisms has increased worldwide [1]. Only two novel classes of antibiotics to fight drug resistant microorganisms have entered the clinic in the past three decades [2]. Moreover, the antibacterial pipeline is scarce because of the costs associated with the clinical trials and

* **Corresponding author Isa Domingues Serrano:** Centro de Investigação Interdisciplinar em Sanidade Animal (CIISA), Faculdade de Medicina Veterinária da Universidade de Lisboa, Avenida da Universidade Técnica, 1300-477 Lisboa, Portugal; E-mail: isaserrano@gmail.com

Manuela Oliveira and Isa Serrano (Eds)

licensing of the antibiotics [2]. The emergence of multiresistant strains and the temporal and geographical variations in resistance patterns suggest the need to look for alternatives to the classic antimicrobial therapy to decrease the morbidity and mortality caused by these disease causing agents. AMPs may act as lead molecules for the development of novel anticancer agents and antimicrobial agents with a broad-spectrum antimicrobial activity displayed at micromolar concentrations or less, and low probability to generate resistance [3]. However, there are some limitations in the application of antimicrobial peptides that calls into question if they will be the raising stars for tomorrow's clinical practice.

1. WHAT ARE ANTIMICROBIAL PEPTIDES (AMPS)

Most Antimicrobial Peptides (AMPs) are 10–40 residue polypeptides. However, some may have more than 100 amino acids. The amino acid sequence of AMPs is part of the genetic code of virtually all multicellular organisms. They are part of the immune system across eukaryotes as a defense strategy, providing a first barrier against dissemination of a wide range of microorganisms, including bacteria, virus, protozoa, and fungi [4, 5]. The antimicrobial activity is derives from their cationic charge due to the presence of multiple lysine, tryptophan, and arginine residues, a large portion of hydrophobic residues (50% or higher), hydrophobicity, and amphipathicity [6].

A single peptide may present multiple functions according to the concept of promiscuity. They may act over different targets depending on their physical-chemical properties (pure promiscuity) or on their amino acid modification (family promiscuity) [7]. Therefore, AMPs exhibit physiological advantages over other molecules for application in the drug discovery and development process. They are promising drug candidates for the control or treatment of human, animal, and plant (agricultural) diseases, or to be used in combination with conventional therapy.

1.1. Mode of Action

It has been generally accepted that the mode of action of AMPs occurs through membrane permebilization owing to the hydrophobic nature of cellular membranes, and less commonly through non-membrane disruption (non-lytic) [6, 8].

Membrane permebilization can either consist in pore formation in the lipid membrane (barrel stave and toroidal pore models), thinning of the membrane bilayer, membrane dissolution (carpet model), or lipid-peptide domain formation [9]. Membrane is where AMPs action begins. The AMPs membrane, cationic and hydrophobic, initially interacts to the anionic lipid head groups of the outer surface of the microbial cytoplasmic membranes which are rich in lipids such as phosphatidylglycerol or cardiolipin. The difference in the exposed lipid content in the membranes of microbes and host cells is the most used explanation for the selectivity of AMPs for microbes over host cells [10]. Therefore, the binding of AMPs to the membranes of microbes is stronger than to the neutral animal plasma membranes rich in phosphatidylcholine/cholesterol/sphingomyelin [10].

After interaction, the peptide is inserted into the outer leaflet of the membrane, roughly parallel to the bilayer, leading to the displacement of lipids and perturbing the membrane. Following insertion into the bilayer, AMPs dissipate the electrochemical gradient across the microbial membrane. Experimentally, this happens within a few seconds after addition of AMPs [11, 12]. Finally, the disruption of the membrane structure occurs (membrane blebbing, vesiculation, fragmentation, DNA release, and cell aggregation) leading to the disruption of the microbial cell morphology [10]. AMPs should be able to rapidly cross through the bacterial cell walls, crossing the thick proteoglycan layer of the Gram-positive bacteria and the layer of LipoPolySaccharide (LPS) in the outer membrane of Gram-negative bacteria.

In non-membrane disruption, AMPs translocate into the cytoplasm and attack internal cellular targets affecting cellular processes from RNA and DNA synthesis [13] to loss of ATP from actively respiring cells [14].

It is likely that most AMPs show several simultaneous mechanisms of action such as membrane permeabilization and intracellular effects. This would help explain AMPs broad-spectrum activity and the scarcity of bacterial resistance to AMPs [10].

Because of the AMPs' ability to translocate into the cytoplasm, membrane-crossing AMPs have also been used as templates for the development of cell-penetrating peptides (CPP) [15]. The ability of AMPs to accumulate at sites of infection combined with their almost negligible cytotoxicity to host cells make them interesting as targeting vectors for positron emission tomography (PET) imaging of infection [16]. Thus, AMPs can have clinical applications as antibiotics, as precursors of drug transporters [8], and as vectors for PET imaging [16].

1.2. Immune System

AMPs are a diverse collection of gene-encoded defense molecules that, beyond being antimicrobial, participate in several aspects of immune system regulation such as: inhibition of pro-inflammatory responses induced by LPS, chemotaxis activity, expression in acquired immunity modulating the responses of dendritic cells or T cells, and expression in wound healing and growth promotion [6, 17]. All of these functions consequently diminish inflammation and infection [17].

AMPs are expressed in the skin and mucosae normally exposed to the external environment and to microbes [18]. They are produced by different blood cells such as eosinophils, neutrophils, platelets, which are involved in the acute inflammatory response. There are three major AMPs classes in humans: defensins, cathelicidins, and thrombocidins. Defensins and cathelicidins are produced by a variety of blood cell types, whereas thrombocidins are released only from platelets. β-defensins are also produced by epithelial cells [19]. Defensins have a β-sheet structure and three disulfide bridges. They are further classified by their disulfide bridging pattern, distinguishing α- and β- defensins. The only cathelicidin produced in humans is the α-helical LL-37 [20]. AMPs are expressed either constitutively (most α-defensins) or induced (most β-defensins) by inflammation or injury [3, 21]. In contrast, dermcidin, an AMP with a broad-spectrum activity, is exclusively part of the constitutive innate defense of human skin, being secreted by the eccrine sweat glands and not inducible by inflammation or injury of the skin [18].

AMPs will confer protection against infections through alert, activating and amplifying innate and acquired immune responses of the host. However, the expression of AMPs can be ambivalent. The decreased expression of AMPs can increase the host susceptibility to infectious diseases. For example, there is a significant higher risk for developing Crohn's disease when individuals have a lower gene copy number (≤3) of hBD2 AMP [17]. On the other hand, an excess of cathelicidin LL37 leads to an abnormal blood vessel growth and inflammation developing the facial skin disease rosacea [17]. In summary, AMPs display a central role in regulating and balancing the inflammatory responses of the immune system.

1.3. Resistance

Bacteria do not appear to develop resistance to AMPs as easily as with conventional antibiotics [22]. The most consensual explanation is related to the ionic interaction between the cationic AMP and the anionic bacteria membrane when they both meet. Interactions with bacterial components do not involve specific protein binding sites [23]. In order to develop resistance to AMPs, bacteria would have to alter their outer membrane, an almost universal structure among bacteria. In addition, the short time-frame of interaction between microorganism and AMP promotes the rapid death of bacteria and decreases the probability to develop resistance [24]. Nevertheless, there are some bacteria which develop less sensitivity to cationic AMPs. These mechanisms comprise: 1) degradation of AMPs, 2) removal of AMPs from their site of action and change of the AMPs target, 3) trapping, and 4) modulation of AMPs gene expression [20, 25] (Table **1**).

Table 1. Mechanisms of resistance showed by distinct bacteria to some AMPs.

AMPs	**Bacteria**	**Resistance mechanisms**
LLR7	*Staphylococcus aureus*	1. aureolisin
	Enterococcus faecalis	1. gelatinase
	Streptococcus pyogenes	1. SpeB 3. M protein / SIC
	Staphylococcus epidermidis	2. Efflux: VraFG 3. exopolisaccharide

(Table 1) contd.....

AMPs	Bacteria	Resistance mechanisms
Lysozyme	*S. aureus, S. epidermidis, Staphylococcus lugdunensis, E. faecalis, Listeria monocytogenes*	2. Peptidoglycan O-acetylase (OatA)
	Streptococcus pneumoniae, E. faecalis, Streptococcus suis, L. monocytogenes, Bacillus anthracis	2. Peptidoglycan N-acetylglucosamine deacetylase A (PdgA)
	Micobacterium smegmatis	2. N-acetylmuramic acid hydroxylase NamH
Bacitracin	*Streptococcus mutans*	3. Efflux: MbrAB
	S. pneumoniae	3. Efflux: SP0812-SP0813
	S. aureus	3. Efflux: VraDE , VraFG
	Bacillus licheniformis	3. Efflux: BcrAB(C)
	E. faecalis	3. Efflux: BcrAB(D)

Adapted from Nawrocki KL, Crispell EK, McBride SM. Antimicrobial Peptide Resistance Mechanisms of Gram-Positive Bacteria. Antibiotics (Basel) 2014; 3(4): 461-92 [26]. The author acknowledges the source.

The first mechanism is based on the secretion of proteases, such as gingipains from *Porphyromonas gingivalis*, which inactivate AMPs [27]. Cathelicidin LL-37 is a human AMP with potent anti-staphylococcal activity. However, LL-37 is susceptible to the degradation by two major proteinases produced by *Staphylococcus aureus*, aureolysin and V8 protease. As a result, these two proteases contribute to the resistance of *S. aureus* to the innate immune system of humans [28].

The second mechanism is the most frequent and includes efflux pumps to remove AMPs from their action site, and modification of the bacterial cell surface composition. Altering surface composition can be achieved by modifying the bacterial phospholipid bilayer fluidity through the introduction of fatty acids with different chain lengths [29]. This process frequently differs among Gram-positive and Gram-negative bacteria. In *S. aureus* (Gram-positive) AMPs resistance is conferred by the substitution of teichoic acids with D-alanine decreasing the negative net charge of teichoic acids [30], whereas in *Salmonella typhimurium* (Gram-negative) is conferred by the modification of lipid A either by the addition of an aminoarabinose moiety (Pmr system) [31] or by acylation (PagP) [32].

The streptococcal inhibitors of complement (SIC) and the staphylokinase of *S. aureus* (SAK) specifically bind to and interact with AMPs to prevent them from reaching the cytoplasmic membrane, in a process called trapping [20, 25].

The last mechanism is shown by *Shigella flexneri* which subvert the immune system by down-regulating the expression of cathelicidin LL-37 and the human β-denfensin-1, leading to increased bacterial adhesion and invasion.

The bacterial resistance mechanisms against anionic AMPs are less recognized. In *Staphylococcus epidermidis* the cationic exopolymer Polysaccharide Intercellular Adhesin (PIA) and the anionic Poly-γ-Glutamic Acid (PGA) apparently protect the bacteria from cationic and anionic AMPs [33]. PIA and alginate are typical components of the bacterial biofilms extracellular matrix, which may explain their higher resistance to AMPs [20].

Moreover, AMPs resistance mechanisms may be activated by bacterial specific sensors, such as the Gram-negative PhoP/PhoQ and the Gram-positive Aps, which include a classical bacterial two-component sensor/regulator system. They were first described in *S. typhimurium* and *S. epidermidis*, respectively, and show significant differences at structural, mechanistic, and functional level. Both mechanisms are crucial for virulence and are potential targets for antimicrobial therapy [20].

An interesting fact is that the physiological concentrations of AMPs *in vivo* are much lower than their Minimal Inhibitory Concentration (MIC) required for *in vitro* antimicrobial activity against bacteria and fungi [17], usually in the 1-50 μg / ml range for most peptides [6]. One possible explanation already demonstrated for a number of AMPs is that these peptides may act synergistically with other AMPs *in vivo* potentiating their antimicrobial efficiency [34].

Despite promising applications of AMPs some researchers fear that their therapeutic use may lead to a directional selection for resistance strongly influencing its evolution. Recently, Habets *et al.* 2012 [35] described an experimentally developed resistance to pexiganan. Pexiganan is a synthetic AMP from frog magainin, candidate for the treatment of infections in diabetic leg ulcers mostly caused by *S. aureus*. This experimentally developed resistance endowed *S.*

aureus with cross-resistance to human neutrophil defensin-1, compromising natural immunity [35]. These results emphasize the risk of AMPs-based therapy and the need to ensure the safety of long term use of AMPs.

2. CHRONOLOGY OF AMPS DISCOVERY – A BRIEF HISTORY

The identification of AMPs can be traced to the discovery of lysozyme by A. Flemming (1921) which is considered by most authors to be the first report of a peptide with antimicrobial activity involving non-enzymatic mechanisms that are similar to AMPs [3].

The antibacterial activity of nisin was identified in 1928; however its structure was described further later (http://aps.unmc.edu/AP). In 1939, gramicidins were isolated from *Bacillus brevis* and were found to exhibit activity both *in vitro* and *in vivo* against a wide range of Gram-positive bacteria [36, 37]. Gramicidin A, B and C from bacteria, were the first AMPs to be commercially manufactured as antibiotics [38]. In 1947, polymyxin was used to treat resistant Gram-negative superbugs.

The first plant AMP, viscotoxin A3, was discovered in 1968. Subsequently AMPs were identified in a large variety of organisms, such as amphibians (bombinin, 1970), insects (melittin A, 1973), and humans (human α-defensins (HNP1), 1985).

The discovery of penicillin in 1929 and streptomycin in 1943 mark the beginning of the age of antibiotics. This period, from 1950 to 1960, was the golden age of antibiotics discovery, as 50% of the drugs commercially used today were discovered and developed for therapeutic use after the World War II. [39]. At this time, there was a loss of interest in the natural host antimicrobial peptides that only returned to the spotlight in the early 1960s with the increase of multidrug-resistant microbial pathogens [40]. It is this point in time that some authors consider to be the true origin of AMPs research [3, 41].

In 1957, it was established that phagocytic cells produce intracellular cationic AMPs in response to infection [42], but only in 1996 it was confirmed for the first time that AMPs also played a role in the defense systems of organisms like the fruit fly without an acquired immune system. The authors of the study demonstrated that the deletion of an AMP gene in *Drosophila melanogaster* rendered it susceptible to fungal infections [43]. Since these studies, AMPs have been extensively studied in other organisms without an acquired immune system like plants, insects and other invertebrates [3].

In 1981, Hans Boman identified the silk moth (*Hyalophora cecropia*) cecropins involved in the insect innate immune response, constituting the first major α-helical AMPs to be reported [3, 44]. In 1987, Michael Zasloff discovered magainins, the Hebrew word for "shield", in the African clawed frog *Xenopus laevis* [3, 45]. The human histatin and bovine bactenecin (the first cathelicidin) were discovered in 1988. TAP (the first β-defensin), the human defensin-6 (HD6), and the human cathelicidin LL-37 were discovered later in the 1990s. Kalata B1, which is a prototype plant cyclotide is only established as an AMP in 1999, two decades after being identified.

These pioneering studies ignited the interest of many investigators in these natural host antimicrobial peptides, leading to the purification and characterization of hundreds of new AMPs such as piscidin 1 (first AMP from mast cells, 2001), and the human keratin-derived AMPs (2012), ([46], http://aps.unmc.edu/AP). The number of discovered AMPs has now reached more than 2000. Currently, the AMP database, at http://aps.unmc.edu/AP, contains over 2400 antimicrobial peptides and proteins, 98.4% from living prokaryotic and eukaryotic organisms (Fig. **1**).

2.1. Classification of AMPs

Most AMPs have common biophysical parameters including small molecular size with less than 60 amino acids, cationicity (typically 4+ or 5+ charged [49]), and amphipathic shape in which clusters of hydrophobic and hydrophilic amino acids segregate. The amphipathic shape is probably essential for AMPs activity and for their ability to preserve cells' integrity [50].

Antimicrobial peptides can be classified based on their biological source, biological functions, biosynthesis, molecular properties, molecular targets, and 3D structure [5]:

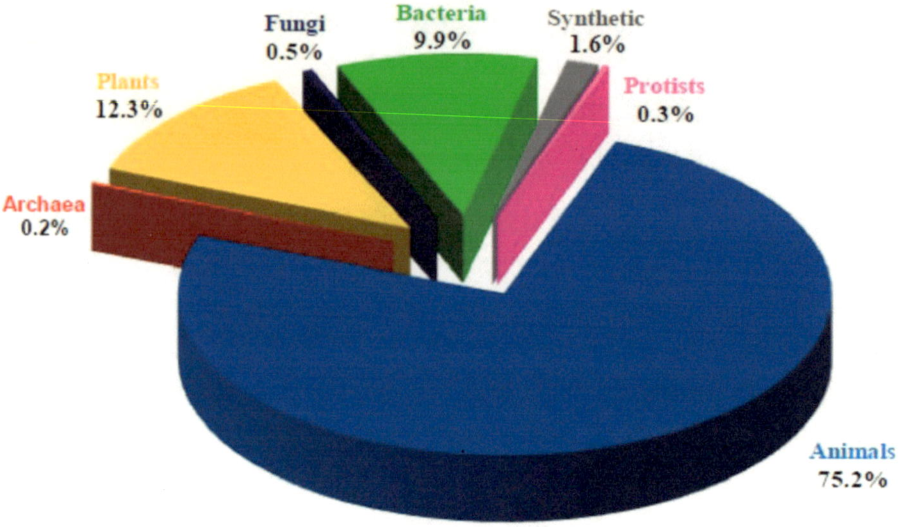

Fig. (1). Sources of AMPs in the antimicrobial peptide database [47, 48] (Total of 2583; July 2015). AMPs mainly derive from animals (75.2%), plants (12.3%), and bacteria (9.9%). Only 2.6% AMPs derive from fungi, protists, archaea and from the synthetic pathway

2.1.1. Biological Source

Bacterial (bacteriocins), plant, and animal including insect, amphibian, fish, reptile, and mammal (cathelicidins, defensins, and histatins) [5].

2.1.2. Biological Functions

Antibacterial, antiviral, antifungal, antiparasital, insecticidal, chemotactic, wound healing, growth promotion.

2.1.3. Biosynthesis

Gene coded (considered by most authors as true AMPs) and non-gene coded (*i.e.* multiple enzyme systems) AMPs [5].

2.1.4. Molecular Properties

Charge (cationic,neutral,and anionic), hydrophobicity (hydrophobic, amphipathic, and hydrophilic), length (ultra-small, small, medium, and large), and motif (linear, loops, circular, α-helical, etc), ([5], http://aps.unmc.edu/AP).

2.1.5. Molecular Targets

Cell surface targeting peptides membrane and non-membrane targeting (*e.g.* nisins and temporins) and intracellular targeting peptides (*e.g.* Pro-rich peptides) [5].

2.1.6. Three-Dimensional (3D) Structure

Based on the types of secondary structures AMPs may exhibit alpha [helical structures (*e.g.* magainins and LL-37)], beta [β-strands (*e.g.* human α-defensins)], alphabeta [helical and β-strands (*e.g.* β-defensins)], and non-alphabeta structures [neither helical nor β-strands (*e.g.* indolicidin)] [5].

An assertive division of AMPs was proposed by Diamond *et al.* 2009 [6]. In this system, AMPs are divided into five categories on the basis of their amino acid composition and structure: 1) anionic, 2) linear amphipathic α-helical, 3) cationic enriched for specific amino acids, 4) peptide fragments, and 5) peptides with cysteines forming intramolecular bonding (Table **2**).

Table 2. Classes of AMPs: their structural properties, source and body location.

Class of AMPs	**Structure**	**Examples, origin and location**
Anionic. Require zinc as a cofactor and display activity against both Gram-positive and Gram-negative bacteria.		Maximin H5 [51]; Dermcidin [52].Amphibians; Humans.Surfactant extracts, bronchoalveolar lavage fluid, and airway epithelial cells.
Linear. ~290. Most have < 40 amino acids in length and possess a 3D structure with a kink or hinge in the middle.	Cationic, α-helical.	Cecropin [44]; Magainin [45]; Pluerocidin [53]; Melittin [54]. Insects; Amphibians; Fish; Insects (bees).Skin.
Linear, Cationic, enriched. ~44. Without cysteine residues, making it very flexible and fluid in solution.	Cationic, enriched in amino acids, extended.	Bactenecins , Proline Arginine-rich peptide (PR-39); Proline rich (38%) abaecin from bees; Indolicidin from cattle, rich in tryptophan [55]. Mammals and insects.Skin, intestinal epithelium.

(Table 2) contd.....

Class of AMPs	Structure	Examples, origin and location
Peptide fragments. Charged peptides that are fragments of larger peptides.	Cationic, β-turn.	Lactoferrin [56]; cathelicidins [57]. Humans.Blood.
Charged peptides with cysteine. ~380 members.	Cationic, anionic, forms intramolecular disulfide bonds and β-sheets, possess Cysteine.	α-, β-, and θ-defensins [50]; Protegrin [58]. Birds, reptiles, mammals, insects, and plants; Mammals (porcine).Ubiquitous.

Adapted from Diamond G, Beckloff N, Weinberg A, *et al.* The roles of antimicrobial peptides in innate host defense. Curr Pharm Des 2009; 15(21): 2377-92 [6]. The author acknowledges the source.

Other peptides, such as the antifungal histatins and the antiviral zap proteins, are not included in any AMPs classes because they are characterized by diverse structures, demonstrating the wide variety of antimicrobial host defense peptides [6]. Some peptides are species-specific and exhibit strain-specific activity, for example the human defensins HNP1-3, hBD1, hBD2, hBD3, and the human cathelicidin LL-37 exhibit different MICs in specific bacterial and fungal strains [6].

2.2. Currently Active AMPs Databases

Databases containing genomic, proteomic and functional information have become crucial in antimicrobial drug research. The first antimicrobial sequence database was placed online in 1997 by Alex Tossi (http://www.bbcm.univ.trieste.it/tossi/pag1.htm, now inactive). After 18 years there are several active databases with distinct functions (Table **3**).

Table 3. Active databases of peptides.

Name	Link	Description of the database
AMPer	http://marray.cmdr.ubc.ca/cgi-bin/amp.pl	Database and program searching for fragments with sequences similar to AMPs [59].

(Table 3) contd.....

Name	Link	Description of the database
APD	http://aps.unmc.edu/AP/main.php	Antimicrobial and anticancer peptides [47, 48].
ArachnoServer	http://www.arachnoserver.org/mainMenu.html	Toxic peptides and proteins from spider venoms [60].
ATDB	http://protchem.hunnu.edu.cn/toxin	Database of toxic proteins and peptides of animal origin [61].
BACTIBASE	http://bactibase.pfba-lab-tun.org	Bacteriocins [62].
BAGEL2	http://bagel2.molgenrug.nl/	Bacteriocins.(University of Groningen, the Netherlands).
CAMP	http://www.bicnirrh.res.in/antimicrobial/	AMPs and proteins [63, 64].
CyBase	http://www.cybase.org.au/	Cyclotides AMPs [65, 66].
DAMPD	http://apps.sanbi.ac.za/dampd/	An updated ANTIMIC database. Enables to search the similarities using BLAST [67].
DAPD	http://split4.pmfst.hr/dadp/	Anuran defense peptides [68].
Defensins Knowledge base	http://defensins.bii.a-star.edu.sg/	Defensins AMPs. (Bioinformatics Institute and Singapore Eye Research Institute, Singapore).

(Table 3) contd.....

Name	Link	Description of the database
Enzybase	http://biotechlab.fudan.edu.cn/database/EnzyBase/home.php	Cell wall-degrading enzymes including lysins, bacteriocins, autolysins, and lysozymes AMPs [69].
EROP-Moscow	http://erop.inbi.ras.ru	Biologically active peptides.(Institute of Biochemistry, Russian Academy of Sciences, Moscow, Russia).
Hmrbase	http://crdd.osdd.net/raghava/hmrbase/	Database of hormones and their receptors [70] .
LAMP	http://biotechlab.fudan.edu.cn/database/lamp/	Database linking AMPs [71].
NORINE	http://bioinfo.lifl.fr/norine/	Nonribosomally synthesized bioactive peptides.(Laboratoire d'informatique fondamentale de Lille, Institut national de recherche en informatique et en automatique, and Laboratoire des procédés biologiques génie enzymatique et microbien, France).
Penbase	http://penbase.immunaqua.com	Shrimp, Litopenaeus vannamei, AMPs [72].
PepBank	http://pepbank.mgh.harvard.edu	A database of biologically active peptides and program searching for fragments with sequences similar to peptides.(Massachusetts General Hospital, Harvard University, USA).

(Table 3) contd.....

Name	Link	Description of the database
Peptaibol	http://peptaibol.cryst.bbk.ac.uk/home.shtml	Fungal AMPs.(School of crystallography, Birkbeck College, Univ. of London, UK).
PeptideDB	http://www.peptides.be/	Major database of biologically active peptides, peptide precursors and motifs in Metazoa.(Data analysis & modeling group at Hasselt University and Functional genomics and proteomics unit at K.U. Leuven, Belgium).
PhytAMP	http://phytamp.pfba-lab-tun.org/main.php	AMPs and proteins of plant origin [73].
RAPD	http://faculty.ist.unomaha.edu/chen/rapd/index.php	Recombinant AMPs [74].
SwePep	http://www.swepep.org	Database of biologically active peptides including sophisticated computing tools for peptide mass spectra interpretation.(Laboratory for biological and medical mass spectrometry at Uppsala University, Sweden).
Thiobase	http://db-mml.sjtu.edu.cn/THIOBASE/	Thiopeptides (sulfur-rich, highly modified heterocyclic peptide antibiotics).(Microbial bioinformatics group at the state key laboratory of microbial metabolism, Shanghai Jiaotong University, China).

(Table 3) contd.....

Name	Link	Description of the database
YADAMP	http://yadamp.unisa.it/	AMPs database. (Soft matter group at the University of Salerno, Italy).

Adapted from Wang G. Antimicrobial Peptides: Discovery, Design and Novel Therapeutic Strategies: CABI; 2010 [5] and Hammami R, Fliss I. Current trends in antimicrobial agent research: chemo- and bioinformatics approaches. Drug Discov Today 2010; 15(13-14): 540-6 [75]. The author acknowledges the sources.

The databases APD, CAMP, or YADAMP all cover AMP sequences from different origins. Some databases focus on AMPs produced by spider (Arachnoserver), bacteria (BACTIBASE), or anurans (DAPD), *whereas* others target specific families of AMPs including defensins (BACTIBASE, BAGEL2, Defensins Knowledge base), or cyclotides (CyBase) (Table **4**).

Table 4. Some prediction servers of AMPs.

Server	Description and link (country, date)
APD2	Database-based classification and prediction of AMPs (USA, 2004). http://aps.unmc.edu/AP/prediction/prediction_main.php
AntiBP	SVM, QM, or ANN prediction of AMPs (India, 2007). http://www.imtech.res.in/raghava/antibp2/
AMPer	An automated discovery tool for gene-coded antimicrobial peptides (Canada, 2007). http://marray.cmdr.ubc.ca/cgi-bin/amp.pl
CAMP	SVM, RF, ANN, or DA prediction of AMPs (India, 2010). http://www.camp.bicnirrh.res.in/predict/
AMPA	Antimicrobial sequence scanning system to predict potential AMP region from proteins (Spain, 2012). http://tcoffee.crg.cat/apps/ampa/do
AVPpred	SVM Prediction of antiviral peptides (China, 2012). http://crdd.osdd.net/servers/avppred/
iAMP-2L	Two-layer Prediction of AMPs (India / USA, 2013). http://www.jci-bioinfo.cn/iAMP-2L

Adapted from Wang G. Antimicrobial Peptides: Discovery, Design and Novel

Therapeutic Strategies: CABI; 2010 [5]. The author acknowledges the source.

Some databases collects only active peptides (EROP-Moscow, NORINE, PepBank, SwePep), others contain a mixture of active and precursor proteins, *e.g.* PeptideDB. They mostly contain proteomic data, whereas CyBase and Pen Base contain both genomic and proteomic data.

Most databases act as simple repositories of AMPs data with different degrees of curation. In opposition, few databases contain tools for sequence analysis which facilitate drug discovery process [75]. Among those, AMPer database, based on hidden Markov models and the SwissProt database could be considered as a tool for the identification of new AMPs (Table **4**).

Other databases such as DAMPD enables to search the similarities using BLAST, PepBank has a program searching for fragments with sequences similar to peptides, and SwePep display sophisticated computing tools for peptide mass spectra interpretation.

The rapid prediction of AMPs and the prediction of the target organisms of those peptides allowed a better management of AMPs use in both medicine and in food sectors. Databases have also contributed to the creation of models which pretend to describe and predict the membrane interactive potential of AMPs, and to describe evolutionary relationships based on phylogenetic analysis of the peptides. The information collected from databases may provide insights into other membrane–peptide interactions [3].

3. THERAPEUTIC APPLICATIONS OF AMPS

Despite the attractive attributes of AMPs, to bring a successful AMP into the market remains a difficult task. The drug development and discovery process is time consuming (up to 10 years) and expensive (millions of dollars) [76]. Production costs are estimated to be approximately \$50-\$400 per gram of amino acid [77]. The reduction of the peptide size maintaining its activity is one plausible solution to decrease the production costs of AMPs [78]. Pharmaceutical companies are available to develop AMPs drugs only if they obey to some requirements: the peptides are effective and safe in clinical trials as required for

FDA approval; its production by chemical or biological means and package is economical; they have an appropriate shelf life; and are profitable comparing to the antibiotics currently available [79]. According to [23], a major bottleneck in the development of new AMPs derives from the inability to fully understand their mechanism of action in physical-chemical terms that are crucial for the peptide design.

Within thousands of natural peptides and millions of synthetic peptides candidates relatively few AMPs have preceded into clinical trials, and those which proceeded none has obtained FDA approval. Those include AMPs indicated for systemic applications: Neuprex (rBPI21); and for topical skin and local applications: Pexiganan (MSI-78), Iseganan (IB-367), Omiganan (MBI-226), MBI 594AN, Histatin Variants, XMP.629 [79]. In order to facilitate the commercial development of AMPs researchers should focus on small peptides. Short AMPs molecules which are now being developed include A4W-GGN5N11, hLF1-11, (CKPV)2, P-113, and LKW [78].

3.1. Anticancer AMPs

According to the International Agency for Research on Cancer (IARC) at 2008 12.7 million of new cancer cases and 7.6 million deaths from cancer occurred worldwide [80], more frequently from lung, breast and colorectal cancers [9, 80]. The current therapies available for cancer, including surgery and chemotherapy, have a relatively low success rate and reveal a high risk of reoccurrence. Furthermore, chemotherapeutic agents target not only malignant cells but all rapidly dividing cells. Consequently, the development of novel therapies that lack the toxicity of conventional chemotherapeutic agents, demonstrating higher cell selectivity and being unaffected by chemoresistance are urgent.

Those AMPs that are able to kill malignant cells can be placed into two groups: a) AMPs highly potent against bacteria and cancer cells but not against normal mammalian cells, and b) AMPs cytotoxic for bacteria, cancer cells, and normal mammalian cells [81]. The former AMPs are promising candidates for cancer therapy and / or treatment. Several anticancer peptides have been developed against human solid tumors (Table **5**) and human hematological tumors, as well as non-human tumors [9].

Table 5. AMPs with activity against human solid tumors, their target cells and mode of action (selected AMPs from those discovered since 2010).

AMP	Target Cells	Mode of Action
NRC-03, NRC-07	Breast	Cell membrane lysis with possible pore formation in mitochondria and ROS production [82].
ERα17p	Breast	Apoptosis induction and massive necrosis [83, 84].
Temporin-1CEa	Breast	Membrane disruption, calcium release, ROS production [85, 86].
CR1166	Breast and pancreas	Apoptosis induction [87].
Dermaseptin B2	Breast and prostate	Necrosis [88].
TfR-lytic peptide	Breast and prostate, gliobastoma, pancreas and bile-duct	Apoptosis induction [89].
A-8R	Prostate	ROS generation and DNA damage [90].
BEPT II and BEPT II-1 from *Bullacta exarata*	Prostate	Apoptosis induction [91].
9 somatostatin peptide analogues	Colon	DNA polymerase β inhibition [92].
Peptide aptamers	Cervix	Apoptosis induction by inhibition of HSP-70 [93].
BPC96	Cervix	Apoptosis induction [94].
MPI-1	Cervix, prostate and hepatocellular adenocarcinoma	Necrosis after cell membrane targeting [95].
Magainin II (MG2A)	Cervix and lung, melanoma	Both necrosis and apoptosis [96].
A(9)K	Cervix, kidney	Both necrosis and apoptosis [97].
SVS-1	Lung, epidermis and breast	Cell membrane disruption *via* pore formation [98].
Myristoyl-Cys-Ala-Val-Ala-Tyr-(1,3 dimethyl)His-OMe	Several cell lines (Lung, colon, breast, ovarian, renal, etc)	DNA synthesis/replication inhibition [99].

Adapted from Gaspar D, Veiga AS, Castanho MA. From antimicrobial to anticancer peptides. A review. Front Microbiol 2013; 4: 294 [9]. The author acknowledges the source.

Some AMPs exert their effect against acute T cell leukemia (Polycationic peptides) [100], chronic myelogenous leukemia and acute lymphoblastic T-leukemia cells (Pep2 and Pep3) [101], histiocytic lymphoma, chronic myelogenous leukemia and acute myeloid leukemia (BIM SAHBA) [102].

The anticancer activity of AMPs can be either membranolytic or non-membranolytic. In membranolytic AMPs the differences between cancer and mammalian normal cells modulate the AMPs' selectivity and toxixity. These differences comprise cell membrane composition, fluidity, and surface area [103]. The membranes of normal cells have mostly zwitterionic phospholipids (neutral net charge) such as phosphatidylethanolamine (PE), phosphatidylcholine (PC) or sphingomyelin (SM), and are rich in cholesterol. These properties probably modulate the cell fluidity leading to the blockage of cationic peptides [81]. In opposition, cancer cells have a net negative membrane charge due to the presence of anionic molecules such as phospholipid phosphatidylserine (PS) (< 9% of the total phospholipids of membranes),O-glycosylated mucins, sialylated gangliosides, heparin [81, 103], and are more fluid [104]. Also, cancer cells have a greater cell surface area owing to the presence of more microvilli, which may increase bonding between the negatively charged membrane lipids of cancerous cells and the positively charged AMPs [81].

Anticancer AMPs exhibit low selectivity and presently none is being commercialized [105]. Another reason behind the difficulty for AMPs enter into clinical trials is related to the use of AMPs with similar sequences to natural AMPs, which may compromise the human immune system and threat public health [106].

To obtain successful AMPs is rather complex. AMPs' physicochemical properties must be manipulated in order to obtain a balanced conjugation of their amino acid sequence, membrane composition, net charge, structural folding in membranes (includes secondary structure, dynamics and orientation), oligomerization capacity, amphipathicity, hydrophobicity, peptide concentration, and high serum stability [9, 81]. The ideal conjugation of these properties will result in AMPs with killing capacity and higher specificity (low cytolytic activity comparing to normal mammalian cells and red blood cells) ready to be used in cancer treatments.

3.2. Antiviral AMPs

The genetic variation of virus, their diversity in the means of transmission, efficient replication within host, and their capacity to persist in the host are obstacles to the control of viral diseases [7]. For that reason, novel antiviral compounds are needed to restrain emerging viral diseases. Since the 1980s, a variety of antiviral peptides from different sources have been isolated (Table **6**).

Table 6. Recent antiviral AMPs: their source, viral target, and mode of action.

Peptide	Source	Target virus	Mode of Action
Caerin 1.1; Caerin 1.9; Maculatin	-	HIV	Disrupts the integrity of the virion membrane [107].
CAP37	*Homo sapiens*	HSV-1, AdV	Disrupts the envelope and/or capsid [108].
HNP-1; HNP-3	*Homo sapiens*	HSV, AdV	Blocks early steps of viral replication [109].
Hp1090	*Heterometrus petersii*	HCV	Disrupts viralmembrane integrity [110].
Lactoferricin	*Homo sapiens* *Bos taurus*	HCMV, HIV, HSV, Papilloma	Activity at virus-cell interface Blocks heparan sulfate [111].
LL-37	*Homo sapiens*	HSV, IAV	Viral receptor-based mechanisms [112].
Pa-MAP	*Pleuronectes americanus*	HSV	Interacts with viral envelope [113].
PAP	*Phytolacca americana*	HIV, HBV, HSV	Inhibits viral protein synthesis [114].

Adapted from Mulder KC, Lima LA, Miranda VJ, et al. Current scenario of peptide-based drugs: the key roles of cationic antitumor and antiviral peptides. Front Microbiol 2013; 4: 321 [7]. The author acknowledges the source.

The first study reporting an antiviral peptide was published in 1986 [115], in which the activity of α-defensin was described as inhibiting several virus, such as herpes simplex virus types 1 and 2 (HSV), and cytomegalovirus (CMV). Interestingly, the same peptide may exert a broad spectrum of antiviral effects and simultaneous exhibit anticancer activity [7].

According to Mulder KC, Lima LA, Miranda VJ, *et al.* [7], more research is needed to unravel the mechanisms of action of antiviral AMPs, to discover new possible targets, and to develop new delivery systems to boost the entry of antiviral AMPs into clinical trials.

3.3. Anti-parasitic AMPs

Research on AMPs against protozoa has been less exhaustive comparing to AMPs against other microorganisms, partially due to the fact that diseases caused by protozoa occur mostly in poor countries of tropical and subtropical areas, and also because most HIV-associated parasitic infections are manageable by current therapies [116]. Nevertheless, to date several AMPs against parasitic diseases have been reported (Table **7**).

Table 7. AMPs against protozoa: their target species and stage, and mode of action.

Peptide	Target protozoan species (stage)a	Mode of Action
Xenopsin; Magainin 2	P. caudatum; T. pyriformis; A. castellanii (tph)	Cell disruption [117].
Bombinin H2; Bombinin H4	Leishmania donovani (pro); Leishmania pifanoi (ama)	Inhibition of cell proliferation [118].
Temporin A; Temporin B	L. donovani (pro); L. pifanoi (ama)	Inhibition of cell proliferation [119].
Buforin 2	C. parvum (spo) (ooc)	Cytotoxic effect on sporozoites [120].
Dermaseptin-H3 (DRS-H3)	L. amazonensis (pro)	Biphasic killing [121].

Adapted from Rivas L, Luque-Ortega JR, Andreu D. Amphibian antimicrobial peptides and Protozoa: lessons from parasites. Biochim Biophys Acta 2009; 1788(8): 1570-81 [116]. The author acknowledges the source.

[a] Tph, trophozoite; Pro, promastigote; Ama, amastigote; Spo, sporozoite; Ooc, oocyst.

Protozoa remain a highly appealing model for AMPs research. A typical protozoan infection reach its final stage after several weeks, whereas bacterial infections occur much rapidly, that is way protozoa have been considered as subverters of the host immune response, where AMPs might be relevant [116].

3.4. New Generation of AMPs

Nowadays, AMPs are designed and engineered for several applications, not only for the prevention of microbial infections but also for food preservation, coatings for implants, catheters, and wound dressings [122]. There are a consistent number of new AMPs classes derived from natural AMPs: a) peptidomimetics, b) congeners, c) conjugates, d) hybrids, e) stabilized AMPs, and f) immobilized AMPs [122].

a. Peptidomimetics are derived from peptoids, β-peptides, arylamides, oligomers or phenylene ethynylenes [122]. Peptidomimetics is the design and synthesis of peptide mimics and it is intend to mimic the structure, function, and mode of action of host natural peptides [123]. It aims to circumvent the AMPs' high susceptibility to microbial proteases, high toxicity and short half-life [123]. Furthermore, antimicrobial peptidomimetics have several advantages over conventional AMPs, such as low immunogenicity, enhanced stability (plasma and metabolic), and cell specificity. Its higher stability results from the using of a peptidomimetic library [123]. However, some questions regarding the hemolytic activity and cytotoxicity of antimicrobial peptidomimetics need to be elucidated in order to ensure their future application *in vivo* [123].
   Recently, a new class of AApeptides peptidomimetics (α-AApeptides and γ-AApeptides) containing N-acylated-N-aminoethyl amino acid units, derived from chiral peptide nucleic acid backbones, have been developed [124].
   There are several antimicrobial peptidomimetics currently in Phase I and II clinical trials [123]:
   - Small arylamide foldamers with increased activity towards *S. aureus* and *Escherichia coli* (Phase II);

- Lytixar TM (also known as LTX-109) which contains a modified tryptophan derivative as lipophilic bulk, displaying high antibacterial activity against methicillin-resistant *Staphylococci* and staphylococcal biofilms (Phase I / IIa);
- Brilacidin (also known as PMX-30063) is a potent bactericidal with broad-spectrum activity against Gram positive (*S. aureus* and *Enterococcus faecium*) and Gram negative (*E. coli*) (Phase II);
- POL7080, Phase I completed. Highly active on multidrug resistant *Pseudomonas* bacteria (opportunistic bacteria causing disease in patients with reduced immunity), with a potent bactericidal activity in a mouse septicemia infection model.

b. AMPs congeners exert either similar or antagonistic effects to an AMP closely related. There are examples of effective congeners of defensins, CAP18, human CAP18/LL-37, SMAP29, and enterocin [122].
c. Conjugation of AMPs is based on the couple of an AMP to a ligand for an improved AMP activity. Examples of possible ligands: an outer surface antibody specific for a receptor (*e.g.* antifungal AMP linked to a Fusarium spp.-specific antibody [125]), micelles and liposome prepared with uptake mediated mechanisms, fatty acids or steroids, and antigens [122]. However, there are a few examples of successful AMPs conjugates.
d. AMPs hybrids are formed by the active regions of two to three natural host antimicrobial peptides. Examples of hybrid AMPs comprise the family of cecropin A–melittin (CEME), CEMA, CP26, and CP29 [122].
e. More stable AMPs may be produced through cyclization. Cyclic conformation increases the resistance to heat and to proteolytic degradation, gathering a higher stability. Circularized peptide analogues of rabbit defensin NP-1 showed a higher stability in adverse environment (higher salt concentration) [122].
f. AMPs are immobilized through their incorporation or adsorption into a variety of materials or surfaces where they still retain their activity against antimicrobials [122]. One example is the oligo-acyl-lysine (OAK) and polylysines. These compounds are peptidomimetics that when are tethered to resins efficiently arrest pathogens in different liquid media. They are used for the detection or depletion of bacteria. [126].

CONCLUDING REMARKS

Regarding the increasing worldwide dissemination of multiresistant micro-organisms, AMPs are strong drug alternatives to the classic antimicrobial therapy to decrease the morbidity and mortality caused by several pathogenic micro-organisms. Since they cannot be administered orally AMPs will never replace conventional antibiotics.

They exhibit a broad-spectrum antimicrobial activity at micromolar concent-rations or less, low probability to generate resistance, and are preferred for topical and injected applications [123]. However, they have been retained in the pharmaceutical pipeline due to some drawbacks, such as lack of oral bioa-vailability, susceptibility to be degraded by proteases, and toxicity due to high amounts of drug needed for therapy (Fig. **2**).

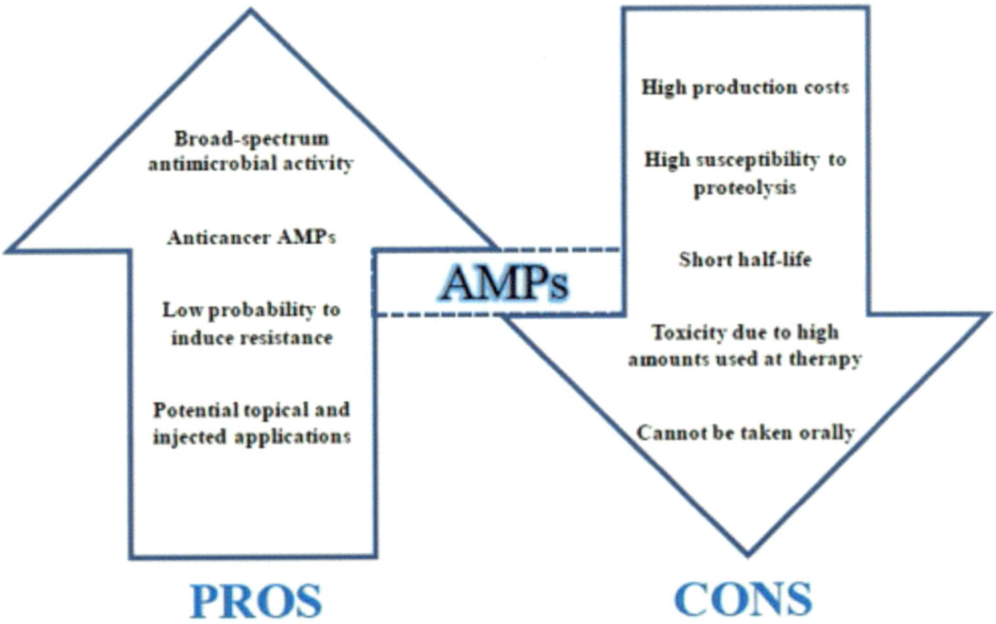

Fig. (2). Key facts for and against AMPs.

Can these faults be circumvented in the future? Understanding the AMPs' fully mechanism of action in physical-chemical terms and investing in the design of the new generation of AMPs are imperative to develop AMPs into useful antimicrobial agents and turn AMPs in the raising stars for tomorrow's clinical practice.

CONFLICT OF INTEREST

The authors confirm that this chapter contents have no conflict of interest.

ACKNOWLEDGEMENTS

The author wish to thank to Professor Manuela Oliveira for the opportunity to write this captivating chapter and to Centro de Investigação Interdisciplinar em Sanidade Animal (CIISA) from Faculdade de Medicina Veterinária da Universidade de Lisboa.

REFERENCES

[1] WHO. Antimicrobial Resistance - Global Report on Surveillance. 2014.

[2] Gupta SK, Nayak RP. Dry antibiotic pipeline: Regulatory bottlenecks and regulatory reforms. J Pharmacol Pharmacother 2014; 5(1): 4-7.
[http://dx.doi.org/10.4103/0976-500X.124405] [PMID: 24554902]

[3] Phoenix DA, Dennison SR, Harris F. Antimicrobial peptides: Their history, evolution, and functional promiscuity. Antimicrobial Peptides. Weinheim, Germany: Wiley-VCH Verlag GmbH & Co KGaA 2013; pp. 1-37.

[4] Toke O. Antimicrobial peptides: new candidates in the fight against bacterial infections. Biopolymers 2005; 80(6): 717-35.
[http://dx.doi.org/10.1002/bip.20286] [PMID: 15880793]

[5] Wang G. Antimicrobial peptides: Discovery design and novel therapeutic strategies: CABI Available from: http://books.google.pt/books?id=QporcUfecJwC. 2010.

[6] Diamond G, Beckloff N, Weinberg A, Kisich KO. The roles of antimicrobial peptides in innate host defense. Curr Pharm Des 2009; 15(21): 2377-92.
[http://dx.doi.org/10.2174/138161209788682325] [PMID: 19601838]

[7] Mulder KC, Lima LA, Miranda VJ, Dias SC, Franco OL. Current scenario of peptide-based drugs: the key roles of cationic antitumor and antiviral peptides. Front Microbiol 2013; 4: 321.
[http://dx.doi.org/10.3389/fmicb.2013.00321] [PMID: 24198814]

[8] Henriques ST, Melo MN, Castanho MA. Cell-penetrating peptides and antimicrobial peptides: how different are they? Biochem J 2006; 399(1): 1-7.
[http://dx.doi.org/10.1042/BJ20061100] [PMID: 16956326]

[9] Gaspar D, Veiga AS, Castanho MA. From antimicrobial to anticancer peptides. A review. Front Microbiol 2013; 4: 294.
[http://dx.doi.org/10.3389/fmicb.2013.00294] [PMID: 24101917]

[10] Wimley WC. Describing the mechanism of antimicrobial peptide action with the interfacial activity model. ACS Chem Biol 2010; 5(10): 905-17.
[http://dx.doi.org/10.1021/cb1001558] [PMID: 20698568]

[11] Matsuzaki K, Sugishita K, Harada M, Fujii N, Miyajima K. Interactions of an antimicrobial peptide, magainin 2, with outer and inner membranes of Gram-negative bacteria. Biochim Biophys Acta 1997; 1327(1): 119-30.
[http://dx.doi.org/10.1016/S0005-2736(97)00051-5] [PMID: 9247173]

[12] Rathinakumar R, Walkenhorst WF, Wimley WC. Broad-spectrum antimicrobial peptides by rational combinatorial design and high-throughput screening: the importance of interfacial activity. J Am Chem Soc 2009; 131(22): 7609-17.
[http://dx.doi.org/10.1021/ja8093247] [PMID: 19445503]

[13] Subbalakshmi C, Sitaram N. Mechanism of antimicrobial action of indolicidin. FEMS Microbiol Lett 1998; 160(1): 91-6.
[http://dx.doi.org/10.1111/j.1574-6968.1998.tb12896.x] [PMID: 9495018]

[14] Kavanagh K, Dowd S. Histatins: antimicrobial peptides with therapeutic potential. J Pharm Pharmacol 2004; 56(3): 285-9.
[http://dx.doi.org/10.1211/0022357022971] [PMID: 15025852]

[15] Magzoub M, Gräslund A. Cell-penetrating peptides: [corrected] from inception to application. Q Rev Biophys 2004; 37(2): 147-95.
[http://dx.doi.org/10.1017/S0033583505004014] [PMID: 15997889]

[16] Ebenhan T, Gheysens O, Kruger HG, *et al.* Antimicrobial peptides: their role as infection-selective tracers for molecular imaging. Biomed Res Int 2014; 2014: 867381.
[http://dx.doi.org/10.1155/2014/867381] [PMID: 25243191]

[17] Lai Y, Gallo RL. AMPed up immunity: how antimicrobial peptides have multiple roles in immune defense. Trends Immunol 2009; 30(3): 131-41.
[http://dx.doi.org/10.1016/j.it.2008.12.003] [PMID: 19217824]

[18] Wiesner J, Vilcinskas A. Antimicrobial peptides: the ancient arm of the human immune system. Virulence 2010; 1(5): 440-64.
[http://dx.doi.org/10.4161/viru.1.5.12983] [PMID: 21178486]

[19] Lehrer RI. Primate defensins. Nat Rev Microbiol 2004; 2(9): 727-38.
[http://dx.doi.org/10.1038/nrmicro976] [PMID: 15372083]

[20] Otto M. Bacterial sensing of antimicrobial peptides. Contrib Microbiol 2009; 16: 136-49.
[http://dx.doi.org/10.1159/000219377] [PMID: 19494583]

[21] Malik A, Batra JK. Antimicrobial activity of human eosinophil granule proteins: involvement in host defence against pathogens. Crit Rev Microbiol 2012; 38(2): 168-81.
[http://dx.doi.org/10.3109/1040841X.2011.645519] [PMID: 22239733]

[22] Yeaman MR, Yount NY. Mechanisms of antimicrobial peptide action and resistance. Pharmacol Rev 2003; 55(1): 27-55.
[http://dx.doi.org/10.1124/pr.55.1.2] [PMID: 12615953]

[23] Wimley WC, Hristova K. Antimicrobial peptides: successes, challenges and unanswered questions. J Membr Biol 2011; 239(1-2): 27-34.
[http://dx.doi.org/10.1007/s00232-011-9343-0] [PMID: 21225255]

[24] Fernebro J. Fighting bacterial infections-future treatment options. Drug Resist Updat 2011; 14(2): 125-39.
[http://dx.doi.org/10.1016/j.drup.2011.02.001] [PMID: 21367651]

[25] Guilhelmelli F, Vilela N, Albuquerque P, Derengowski LdaS, Silva-Pereira I, Kyaw CM. Antibiotic development challenges: the various mechanisms of action of antimicrobial peptides and of bacterial resistance. Front Microbiol 2013; 4: 353.
[http://dx.doi.org/10.3389/fmicb.2013.00353] [PMID: 24367355]

[26] Nawrocki KL, Crispell EK, McBride SM. Antimicrobial peptide resistance mechanisms of gram-positive bacteria. Antibiotics (Basel) 2014; 3(4): 461-92.
[http://dx.doi.org/10.3390/antibiotics3040461] [PMID: 25419466]

[27] Devine DA, Marsh PD, Percival RS, Rangarajan M, Curtis MA. Modulation of antibacterial peptide activity by products of Porphyromonas gingivalis and Prevotella spp. Microbiology 1999; 145(Pt 4): 965-71.
[http://dx.doi.org/10.1099/13500872-145-4-965] [PMID: 10220176]

[28] Sieprawska-Lupa M, Mydel P, Krawczyk K, *et al.* Degradation of human antimicrobial peptide LL-37 by *Staphylococcus aureus*-derived proteinases. Antimicrob Agents Chemother 2004; 48(12): 4673-9.
[http://dx.doi.org/10.1128/AAC.48.12.4673-4679.2004] [PMID: 15561843]

[29] Jing W, Prenner EJ, Vogel HJ, Waring AJ, Lehrer RI, Lohner K. Headgroup structure and fatty acid chain length of the acidic phospholipids modulate the interaction of membrane mimetic vesicles with the antimicrobial peptide protegrin-1. J Pept Sci 2005; 11(11): 735-43.
[http://dx.doi.org/10.1002/psc.702] [PMID: 16059971]

[30] Peschel A, Otto M, Jack RW, Kalbacher H, Jung G, Götz F. Inactivation of the dlt operon in Staphylococcus aureus confers sensitivity to defensins, protegrins, and other antimicrobial peptides. J Biol Chem 1999; 274(13): 8405-10.
[http://dx.doi.org/10.1074/jbc.274.13.8405] [PMID: 10085071]

[31] Gunn JS, Lim KB, Krueger J, *et al.* PmrA-PmrB-regulated genes necessary for 4-aminoarabinose lipid A modification and polymyxin resistance. Mol Microbiol 1998; 27(6): 1171-82.
[http://dx.doi.org/10.1046/j.1365-2958.1998.00757.x] [PMID: 9570402]

[32] Guo L, Lim KB, Poduje CM, *et al.* Lipid A acylation and bacterial resistance against vertebrate antimicrobial peptides. Cell 1998; 95(2): 189-98.
[http://dx.doi.org/10.1016/S0092-8674(00)81750-X] [PMID: 9790526]

[33] Kocianova S, Vuong C, Yao Y, *et al.* Key role of poly-gamma-DL-glutamic acid in immune evasion and virulence of Staphylococcus epidermidis. J Clin Invest 2005; 115(3): 688-94.
[http://dx.doi.org/10.1172/JCI200523523] [PMID: 15696197]

[34] Cassone M, Otvos L Jr. Synergy among antibacterial peptides and between peptides and small-molecule antibiotics. Expert Rev Anti Infect Ther 2010; 8(6): 703-16.
[http://dx.doi.org/10.1586/eri.10.38] [PMID: 20521897]

[35] Habets MG, Brockhurst MA. Therapeutic antimicrobial peptides may compromise natural immunity. Biol Lett 2012; 8(3): 416-8.
[http://dx.doi.org/10.1098/rsbl.2011.1203] [PMID: 22279153]

[36] Dubos RJ. Studies on a Bactericidal Agent Extracted from a Soil Bacillus: Ii. Protective Effect of the Bactericidal Agent against Experimental Pneumococcus Infections in Mice. J Exp Med 1939; 70(1): 11-7.
[http://dx.doi.org/10.1084/jem.70.1.11] [PMID: 19870886]

[37] Dubos RJ. Studies on a Bactericidal Agent Extracted from a Soil Bacillus: I. Preparation of the Agent. Its Activity *in Vitro.* J Exp Med 1939; 70(1): 1-10.
[http://dx.doi.org/10.1084/jem.70.1.1] [PMID: 19870884]

[38] Van Epps HL. René Dubos: unearthing antibiotics. J Exp Med 2006; 203(2): 259.
[http://dx.doi.org/10.1084/jem.2032fta] [PMID: 16528813]

[39] Davies J. Where have All the Antibiotics Gone? Can J Infect Dis Med Microbiol 2006; 17(5): 287-90.
[PMID: 18382641]

[40] Katz ML, Mueller LV, Polyakov M, Weinstock SF. Where have all the antibiotic patents gone? Nat Biotechnol 2006; 24(12): 1529-31.
[http://dx.doi.org/10.1038/nbt1206-1529] [PMID: 17160057]

[41] Nakatsuji T, Gallo RL. Antimicrobial peptides: old molecules with new ideas. J Invest Dermatol 2012; 132(3 Pt 2): 887-95.
[http://dx.doi.org/10.1038/jid.2011.387] [PMID: 22158560]

[42] Skarnes RC, Watson DW. Antimicrobial factors of normal tissues and fluids. Bacteriol Rev 1957; 21(4): 273-94.
[PMID: 13488885]

[43] Lemaitre B, Nicolas E, Michaut L, Reichhart JM, Hoffmann JA. The dorsoventral regulatory gene cassette spätzle/Toll/cactus controls the potent antifungal response in Drosophila adults. Cell 1996; 86(6): 973-83.
[http://dx.doi.org/10.1016/S0092-8674(00)80172-5] [PMID: 8808632]

[44] Steiner H, Hultmark D, Engström A, Bennich H, Boman HG. Sequence and specificity of two antibacterial proteins involved in insect immunity. Nature 1981; 292(5820): 246-8.
[http://dx.doi.org/10.1038/292246a0] [PMID: 7019715]

[45] Zasloff M. Magainins, a class of antimicrobial peptides from Xenopus skin: isolation, characterization of two active forms, and partial cDNA sequence of a precursor. Proc Natl Acad Sci USA 1987; 84(15): 5449-53.
[http://dx.doi.org/10.1073/pnas.84.15.5449] [PMID: 3299384]

[46] Wang G. Database-Guided Discovery of Potent Peptides to Combat HIV-1 or Superbugs. Pharmaceuticals (Basel) 2013; 6(6): 728-58.
[http://dx.doi.org/10.3390/ph6060728] [PMID: 24276259]

[47] Wang G, Li X, Wang Z. APD2: the updated antimicrobial peptide database and its application in peptide design. Nucleic Acids Res 2009; 37(Database issue): D933-7.
[http://dx.doi.org/10.1093/nar/gkn823] [PMID: 18957441]

[48] Wang Z, Wang G. APD: the Antimicrobial Peptide Database. Nucleic Acids Res 2004; 32(Database issue): D590-2.
[http://dx.doi.org/10.1093/nar/gkh025] [PMID: 14681488]

[49] Hancock RE. Peptide antibiotics. Lancet 1997; 349(9049): 418-22.
[http://dx.doi.org/10.1016/S0140-6736(97)80051-7] [PMID: 9033483]

[50] Zasloff M. Antimicrobial peptides of multicellular organisms. Nature 2002; 415(6870): 389-95.
[http://dx.doi.org/10.1038/415389a] [PMID: 11807545]

[51] Lai R, Liu H, Hui Lee W, Zhang Y. An anionic antimicrobial peptide from toad Bombina maxima. Biochem Biophys Res Commun 2002; 295(4): 796-9.
[http://dx.doi.org/10.1016/S0006-291X(02)00762-3] [PMID: 12127963]

[52] Schittek B, Hipfel R, Sauer B, *et al.* Dermcidin: a novel human antibiotic peptide secreted by sweat glands. Nat Immunol 2001; 2(12): 1133-7.
[http://dx.doi.org/10.1038/ni732] [PMID: 11694882]

[53] Cole AM, Weis P, Diamond G. Isolation and characterization of pleurocidin, an antimicrobial peptide in the skin secretions of winter flounder. J Biol Chem 1997; 272(18): 12008-13.
[http://dx.doi.org/10.1074/jbc.272.18.12008] [PMID: 9115266]

[54] Fennell JF, Shipman WH, Cole LJ. Antibacterial action of melittin, a polypeptide from bee venom. Proc Soc Exp Biol Med 1968; 127(3): 707-10.
[http://dx.doi.org/10.3181/00379727-127-32779] [PMID: 4870538]

[55] Brogden KA. Antimicrobial peptides: pore formers or metabolic inhibitors in bacteria? Nat Rev Microbiol 2005; 3(3): 238-50.
[http://dx.doi.org/10.1038/nrmicro1098] [PMID: 15703760]

[56] Bellamy W, Takase M, Wakabayashi H, Kawase K, Tomita M. Antibacterial spectrum of lactoferricin B, a potent bactericidal peptide derived from the N-terminal region of bovine lactoferrin. J Appl Bacteriol 1992; 73(6): 472-9.
[http://dx.doi.org/10.1111/j.1365-2672.1992.tb05007.x] [PMID: 1490908]

[57] Zanetti M, Gennaro R, Romeo D. Cathelicidins: a novel protein family with a common proregion and a variable C-terminal antimicrobial domain. FEBS Lett 1995; 374(1): 1-5.
[http://dx.doi.org/10.1016/0014-5793(95)01050-O] [PMID: 7589491]

[58] Steinberg DA, Hurst MA, Fujii CA, *et al.* Protegrin-1: a broad-spectrum, rapidly microbicidal peptide with *in vivo* activity. Antimicrob Agents Chemother 1997; 41(8): 1738-42.
[PMID: 9257752]

[59] Fjell CD, Hancock RE, Cherkasov A. AMPer: a database and an automated discovery tool for antimicrobial peptides. Bioinformatics 2007; 23(9): 1148-55.
[http://dx.doi.org/10.1093/bioinformatics/btm068] [PMID: 17341497]

[60] Herzig V, Wood DL, Newell F, *et al.* ArachnoServer 2.0, an updated online resource for spider toxin sequences and structures. Nucleic Acids Res 2011; 39(Database issue): D653-7.
[http://dx.doi.org/10.1093/nar/gkq1058] [PMID: 21036864]

[61] He QY, He QZ, Deng XC, *et al.* ATDB: a uni-database platform for animal toxins. Nucleic Acids Res 2008; 36(Database issue): D293-7.

[PMID: 17933766]

[62] Hammami R, Zouhir A, Ben Hamida J, Fliss I. BACTIBASE: a new web-accessible database for bacteriocin characterization. BMC Microbiol 2007; 7: 89.
[http://dx.doi.org/10.1186/1471-2180-7-89] [PMID: 17941971]

[63] Thomas S, Karnik S, Barai RS, Jayaraman VK, Idicula-Thomas S. CAMP: a useful resource for research on antimicrobial peptides. Nucleic Acids Res 2010; 38(Database issue): D774-80.
[http://dx.doi.org/10.1093/nar/gkp1021] [PMID: 19923233]

[64] Waghu FH, Gopi L, Barai RS, Ramteke P, Nizami B, Idicula-Thomas S. CAMP: Collection of sequences and structures of antimicrobial peptides. Nucleic Acids Res 2014; 42(Database issue): D1154-8.
[http://dx.doi.org/10.1093/nar/gkt1157] [PMID: 24265220]

[65] Mulvenna JP, Wang C, Craik DJ. CyBase: a database of cyclic protein sequence and structure. Nucleic Acids Res 2006; 34(Database issue): D192-4.
[http://dx.doi.org/10.1093/nar/gkj005] [PMID: 16381843]

[66] Wang CK, Kaas Q, Chiche L, Craik DJ. CyBase: a database of cyclic protein sequences and structures, with applications in protein discovery and engineering. Nucleic Acids Res 2008; 36(Database issue): D206-10.
[PMID: 17986451]

[67] Seshadri Sundararajan V, Gabere MN, Pretorius A, *et al.* DAMPD: a manually curated antimicrobial peptide database. Nucleic Acids Res 2012; 40(Database issue): D1108-12.
[http://dx.doi.org/10.1093/nar/gkr1063] [PMID: 22110032]

[68] Novković M, Simunić J, Bojović V, Tossi A, Juretić D. DADP: the database of anuran defense peptides. Bioinformatics 2012; 28(10): 1406-7.
[http://dx.doi.org/10.1093/bioinformatics/bts141] [PMID: 22467909]

[69] Wu H, Lu H, Huang J, Li G, Huang Q. EnzyBase: a novel database for enzybiotic studies. BMC Microbiol 2012; 12: 54.
[http://dx.doi.org/10.1186/1471-2180-12-54] [PMID: 22489867]

[70] Rashid M, Singla D, Sharma A, Kumar M, Raghava GP. Hmrbase: a database of hormones and their receptors. BMC Genomics 2009; 10: 307.
[http://dx.doi.org/10.1186/1471-2164-10-307] [PMID: 19589147]

[71] Zhao X, Wu H, Lu H, Li G, Huang Q. LAMP: A Database Linking Antimicrobial Peptides. PLoS One 2013; 8(6): e66557.
[http://dx.doi.org/10.1371/journal.pone.0066557] [PMID: 23825543]

[72] Gueguen Y, Garnier J, Robert L, *et al.* PenBase, the shrimp antimicrobial peptide penaeidin database: sequence-based classification and recommended nomenclature. Dev Comp Immunol 2006; 30(3): 283-8.
[http://dx.doi.org/10.1016/j.dci.2005.04.003] [PMID: 15963564]

[73] Hammami R, Ben Hamida J, Vergoten G, Fliss I. PhytAMP: a database dedicated to antimicrobial plant peptides. Nucleic Acids Res 2009; 37(Database issue): D963-8.
[http://dx.doi.org/10.1093/nar/gkn655] [PMID: 18836196]

[74] Li Y, Chen Z. RAPD: a database of recombinantly-produced antimicrobial peptides. FEMS Microbiol Lett 2008; 289(2): 126-9.
[http://dx.doi.org/10.1111/j.1574-6968.2008.01357.x] [PMID: 19054102]

[75] Hammami R, Fliss I. Current trends in antimicrobial agent research: chemo- and bioinformatics approaches. Drug Discov Today 2010; 15(13-14): 540-6.
[http://dx.doi.org/10.1016/j.drudis.2010.05.002] [PMID: 20546918]

[76] LaMattina JL. Drug Truths: Dispelling the Myths about Pharma R& D. Hoboken, New Jersey: John Wiley & Sons, Inc 2009.

[77] Marr AK, Gooderham WJ, Hancock RE. Antibacterial peptides for therapeutic use: obstacles and realistic outlook. Curr Opin Pharmacol 2006; 6(5): 468-72.
[http://dx.doi.org/10.1016/j.coph.2006.04.006] [PMID: 16890021]

[78] Seo MD, Won HS, Kim JH, Mishig-Ochir T, Lee BJ. Antimicrobial peptides for therapeutic applications: a review. Molecules 2012; 17(10): 12276-86.
[http://dx.doi.org/10.3390/molecules171012276] [PMID: 23079498]

[79] Gordon YJ, Romanowski EG, McDermott AM. A review of antimicrobial peptides and their therapeutic potential as anti-infective drugs. Curr Eye Res 2005; 30(7): 505-15.
[http://dx.doi.org/10.1080/02713680590968637] [PMID: 16020284]

[80] Ferlay J, Shin HR, Bray F, Forman D, Mathers C, Parkin DM. Estimates of worldwide burden of cancer in 2008: GLOBOCAN 2008. Int J Cancer 2010; 127(12): 2893-917.
[http://dx.doi.org/10.1002/ijc.25516] [PMID: 21351269]

[81] Hoskin DW, Ramamoorthy A. Studies on anticancer activities of antimicrobial peptides. Biochim Biophys Acta 2008; 1778(2): 357-75.
[http://dx.doi.org/10.1016/j.bbamem.2007.11.008] [PMID: 18078805]

[82] Hilchie AL, Doucette CD, Pinto DM, Patrzykat A, Douglas S, Hoskin DW. Pleurocidin-family cationic antimicrobial peptides are cytolytic for breast carcinoma cells and prevent growth of tumor xenografts. Breast Cancer Res 2011; 13(5): R102.
[http://dx.doi.org/10.1186/bcr3043] [PMID: 22023734]

[83] Byrne C, Khemtémourian L, Pelekanou V, *et al.* ERα17p, a peptide reproducing the hinge region of the estrogen receptor α associates to biological membranes: A biophysical approach. Steroids 2012; 77(10): 979-87.
[http://dx.doi.org/10.1016/j.steroids.2012.02.022] [PMID: 22426414]

[84] Pelekanou V, Kampa M, Gallo D, *et al.* The estrogen receptor alpha-derived peptide ERα17p (P(295)-T(311)) exerts pro-apoptotic actions in breast cancer cells *in vitro* and *in vivo*, independently from their ERα status. Mol Oncol 2011; 5(1): 36-47.
[http://dx.doi.org/10.1016/j.molonc.2010.11.001] [PMID: 21163714]

[85] Wang C, Tian LL, Li S, *et al.* Rapid cytotoxicity of antimicrobial peptide tempoprin-1CEa in breast cancer cells through membrane destruction and intracellular calcium mechanism. PLoS One 2013; 8(4): e60462.
[http://dx.doi.org/10.1371/journal.pone.0060462] [PMID: 23577112]

[86] Wang C, Zhou Y, Li S, *et al.* Anticancer mechanisms of temporin-1CEa, an amphipathic α-helical

antimicrobial peptide, in Bcap-37 human breast cancer cells. Life Sci 2013; 92(20-21): 1004-14. [http://dx.doi.org/10.1016/j.lfs.2013.03.016] [PMID: 23583573]

[87] Patra CR, Rupasinghe CN, Dutta SK, *et al.* Chemically modified peptides targeting the PDZ domain of GIPC as a therapeutic approach for cancer. ACS Chem Biol 2012; 7(4): 770-9. [http://dx.doi.org/10.1021/cb200536r] [PMID: 22292614]

[88] van Zoggel H, Carpentier G, Dos Santos C, *et al.* Antitumor and angiostatic activities of the antimicrobial peptide dermaseptin B2. PLoS One 2012; 7(9): e44351. [http://dx.doi.org/10.1371/journal.pone.0044351] [PMID: 23028527]

[89] Kawamoto M, Horibe T, Kohno M, Kawakami K. A novel transferrin receptor-targeted hybrid peptide disintegrates cancer cell membrane to induce rapid killing of cancer cells. BMC Cancer 2011; 11: 359. [http://dx.doi.org/10.1186/1471-2407-11-359] [PMID: 21849092]

[90] Gao S, Hsieh CL, Bhansali M, Kannan A, Shemshedini L. A peptide against soluble guanylyl cyclase α1: a new approach to treating prostate cancer. PLoS One 2013; 8(5): e64189. [http://dx.doi.org/10.1371/journal.pone.0064189] [PMID: 23724033]

[91] Ma J, Huang F, Lin H, Wang X. Isolation and purification of a peptide from Bullacta exarata and its impaction of apoptosis on prostate cancer cell. Mar Drugs 2013; 11(1): 266-73. [http://dx.doi.org/10.3390/md11010266] [PMID: 23344115]

[92] Kuriyama I, Miyazaki A, Tsuda Y, Yoshida H, Mizushina Y. Inhibitory effect of novel somatostatin peptide analogues on human cancer cell growth based on the selective inhibition of DNA polymerase β. Bioorg Med Chem 2013; 21(2): 403-11. [http://dx.doi.org/10.1016/j.bmc.2012.11.024] [PMID: 23266186]

[93] Rérole AL, Gobbo J, De Thonel A, *et al.* Peptides and aptamers targeting HSP70: a novel approach for anticancer chemotherapy. Cancer Res 2011; 71(2): 484-95. [http://dx.doi.org/10.1158/0008-5472.CAN-10-1443] [PMID: 21224349]

[94] Feliu L, Oliveras G, Cirac AD, *et al.* Antimicrobial cyclic decapeptides with anticancer activity. Peptides 2010; 31(11): 2017-26. [http://dx.doi.org/10.1016/j.peptides.2010.07.027] [PMID: 20708052]

[95] Zhang W, Li J, Liu LW, *et al.* A novel analog of antimicrobial peptide Polybia-MPI, with thioamide bond substitution, exhibits increased therapeutic efficacy against cancer and diminished toxicity in mice. Peptides 2010; 31(10): 1832-8. [http://dx.doi.org/10.1016/j.peptides.2010.06.019] [PMID: 20600424]

[96] Liu S, Yang H, Wan L, Cheng J, Lu X. Penetratin-mediated delivery enhances the antitumor activity of the cationic antimicrobial peptide Magainin II. Cancer Biother Radiopharm 2013; 28(4): 289-97. [http://dx.doi.org/10.1089/cbr.2012.1328] [PMID: 23286306]

[97] Xu H, Chen CX, Hu J, *et al.* Dual modes of antitumor action of an amphiphilic peptide A(9)K. Biomaterials 2013; 34(11): 2731-7. [http://dx.doi.org/10.1016/j.biomaterials.2012.12.039] [PMID: 23352040]

[98] Gaspar D, Veiga AS, Sinthuvanich C, Schneider JP, Castanho MA. Anticancer peptide SVS-1: efficacy precedes membrane neutralization. Biochemistry 2012; 51(32): 6263-5. [http://dx.doi.org/10.1021/bi300836r] [PMID: 22839778]

[99] Ourth DD. Antitumor cell activity *in vitro* by myristoylated-peptide. Biomed Pharmacother 2011; 65(4): 271-4.
[http://dx.doi.org/10.1016/j.biopha.2011.02.015] [PMID: 21741202]

[100] Lemeshko VV. Electrical potentiation of the membrane permeabilization by new peptides with anticancer properties. Biochim Biophys Acta 2013; 1828(3): 1047-56.
[http://dx.doi.org/10.1016/j.bbamem.2012.12.012] [PMID: 23262194]

[101] Edison N, Reingewertz TH, Gottfried Y, *et al.* Peptides mimicking the unique ARTS-XIAP binding site promote apoptotic cell death in cultured cancer cells. Clin Cancer Res 2012; 18(9): 2569-78.
[http://dx.doi.org/10.1158/1078-0432.CCR-11-1430] [PMID: 22392914]

[102] LaBelle JL, Katz SG, Bird GH, *et al.* A stapled BIM peptide overcomes apoptotic resistance in hematologic cancers. J Clin Invest 2012; 122(6): 2018-31.
[http://dx.doi.org/10.1172/JCI46231] [PMID: 22622039]

[103] Schweizer F. Cationic amphiphilic peptides with cancer-selective toxicity. Eur J Pharmacol 2009; 625(1-3): 190-4.
[http://dx.doi.org/10.1016/j.ejphar.2009.08.043] [PMID: 19835863]

[104] Sok M, Sentjurc M, Schara M. Membrane fluidity characteristics of human lung cancer. Cancer Lett 1999; 139(2): 215-20.
[http://dx.doi.org/10.1016/S0304-3835(99)00044-0] [PMID: 10395181]

[105] Hu J, Chen C, Zhang S, *et al.* Designed antimicrobial and antitumor peptides with high selectivity. Biomacromolecules 2011; 12(11): 3839-43.
[http://dx.doi.org/10.1021/bm201098j] [PMID: 21955251]

[106] Chen C, Hu J, Zhang S, *et al.* Molecular mechanisms of antibacterial and antitumor actions of designed surfactant-like peptides. Biomaterials 2012; 33(2): 592-603.
[http://dx.doi.org/10.1016/j.biomaterials.2011.09.059] [PMID: 21986402]

[107] VanCompernolle SE, Taylor RJ, Oswald-Richter K, *et al.* Antimicrobial peptides from amphibian skin potently inhibit human immunodeficiency virus infection and transfer of virus from dendritic cells to T cells. J Virol 2005; 79(18): 11598-606.
[http://dx.doi.org/10.1128/JVI.79.18.11598-11606.2005] [PMID: 16140737]

[108] Gordon YJ, Romanowski EG, Shanks RM, Yates KA, Hinsley H, Pereira HA. CAP37-derived antimicrobial peptides have *in vitro* antiviral activity against adenovirus and herpes simplex virus type 1. Curr Eye Res 2009; 34(3): 241-9.
[http://dx.doi.org/10.1080/02713680802714066] [PMID: 19274533]

[109] Hook LM, Lubinski JM, Jiang M, Pangburn MK, Friedman HM. Herpes simplex virus type 1 and 2 glycoprotein C prevents complement-mediated neutralization induced by natural immunoglobulin M antibody. J Virol 2006; 80(8): 4038-46.
[http://dx.doi.org/10.1128/JVI.80.8.4038-4046.2006] [PMID: 16571820]

[110] Yan R, Zhao Z, He Y, *et al.* A new natural α-helical peptide from the venom of the scorpion Heterometrus petersii kills HCV. Peptides 2011; 32(1): 11-9.
[http://dx.doi.org/10.1016/j.peptides.2010.10.008] [PMID: 20950663]

[111] Mistry N, Drobni P, Näslund J, Sunkari VG, Jenssen H, Evander M. The anti-papillomavirus activity

of human and bovine lactoferricin. Antiviral Res 2007; 75(3): 258-65.
[http://dx.doi.org/10.1016/j.antiviral.2007.03.012] [PMID: 17481742]

[112] Barlow PG, Svoboda P, Mackellar A, *et al.* Antiviral activity and increased host defense against influenza infection elicited by the human cathelicidin LL-37. PLoS One 2011; 6(10): e25333.
[http://dx.doi.org/10.1371/journal.pone.0025333] [PMID: 22031815]

[113] Migliolo L, Silva ON, Silva PA, *et al.* Structural and functional characterization of a multifunctional alanine-rich peptide analogue from Pleuronectes americanus. PLoS One 2012; 7(10): e47047.
[http://dx.doi.org/10.1371/journal.pone.0047047] [PMID: 23056574]

[114] Kaur I, Gupta RC, Puri M. Ribosome inactivating proteins from plants inhibiting viruses. Virol Sin 2011; 26(6): 357-65.
[http://dx.doi.org/10.1007/s12250-011-3223-8] [PMID: 22160935]

[115] Daher KA, Selsted ME, Lehrer RI. Direct inactivation of viruses by human granulocyte defensins. J Virol 1986; 60(3): 1068-74.
[PMID: 3023659]

[116] Rivas L, Luque-Ortega JR, Andreu D. Amphibian antimicrobial peptides and Protozoa: lessons from parasites. Biochim Biophys Acta 2009; 1788(8): 1570-81.
[http://dx.doi.org/10.1016/j.bbamem.2008.11.002] [PMID: 19046939]

[117] Soravia E, Martini G, Zasloff M. Antimicrobial properties of peptides from Xenopus granular gland secretions. FEBS Lett 1988; 228(2): 337-40.
[http://dx.doi.org/10.1016/0014-5793(88)80027-9] [PMID: 3125066]

[118] Mangoni ML, Papo N, Saugar JM, *et al.* Effect of natural L- to D-amino acid conversion on the organization, membrane binding, and biological function of the antimicrobial peptides bombinins H. Biochemistry 2006; 45(13): 4266-76.
[http://dx.doi.org/10.1021/bi052150y] [PMID: 16566601]

[119] Mangoni ML, Saugar JM, Dellisanti M, Barra D, Simmaco M, Rivas L. Temporins, small antimicrobial peptides with leishmanicidal activity. J Biol Chem 2005; 280(2): 984-90.
[http://dx.doi.org/10.1074/jbc.M410795200] [PMID: 15513914]

[120] Giacometti A, Cirioni O, Del Prete MS, Barchiesi F, Scalise G. Short-term exposure to membrane-active antibiotics inhibits Cryptosporidium parvum infection in cell culture. Antimicrob Agents Chemother 2000; 44(12): 3473-5.
[http://dx.doi.org/10.1128/AAC.44.12.3473-3475.2000] [PMID: 11083662]

[121] Brand GD, Leite JR, de Sá Mandel SM, *et al.* Novel dermaseptins from Phyllomedusa hypochondrialis (Amphibia). Biochem Biophys Res Commun 2006; 347(3): 739-46.
[http://dx.doi.org/10.1016/j.bbrc.2006.06.168] [PMID: 16844081]

[122] Brogden NK, Brogden KA. Will new generations of modified antimicrobial peptides improve their potential as pharmaceuticals? Int J Antimicrob Agents 2011; 38(3): 217-25.
[PMID: 21733662]

[123] Méndez-Samperio P. Peptidomimetics as a new generation of antimicrobial agents: current progress. Infect Drug Resist 2014; 7: 229-37.
[http://dx.doi.org/10.2147/IDR.S49229] [PMID: 25210467]

[124] Ishitsuka Y, Arnt L, Majewski J, *et al.* Amphiphilic poly(phenyleneethynylene)s can mimic antimicrobial peptide membrane disordering effect by membrane insertion. J Am Chem Soc 2006; 128(40): 13123-9.
[http://dx.doi.org/10.1021/ja061186q] [PMID: 17017792]

[125] Peschen D, Li HP, Fischer R, Kreuzaler F, Liao YC. Fusion proteins comprising a Fusarium-specific antibody linked to antifungal peptides protect plants against a fungal pathogen. Nat Biotechnol 2004; 22(6): 732-8.
[http://dx.doi.org/10.1038/nbt970] [PMID: 15146196]

[126] Rotem S, Raz N, Kashi Y, Mor A. Bacterial capture by peptide-mimetic oligoacyllysine surfaces. Appl Environ Microbiol 2010; 76(10): 3301-7.
[http://dx.doi.org/10.1128/AEM.00532-10] [PMID: 20363797]

CHAPTER 4

Probiotics: Ways of Action and Beneficial Effects

Nabil Benomar, Hikmate Abriouel, Rubén Pérez Pulido, Leyre Lavilla Lerma, María del Carmen Casado Muñoz, Antonio Gálvez*

Area de Microbiología, Departamento de Ciencias de la Salud, Facultad de Ciencias Experimentales, Universidad de Jaén, 23071 Jaén, Spain

Abstract: Probiotics are defined by Food and Agriculture Organization (FAO) and World Health Organization (WHO) as: "live microorganisms that, when administered in adequate amounts, confer a health benefit on the host". *Bifidobacterium* and *Lactobacillus* are the principal probiotic groups; however, there are reports on the probiotic potential of *Enterococcus, Bacillus* and yeasts. A number of potential benefits of probiotics have been proposed, including: alleviation of lactose intolerance, immunomodulatory activity, hypocholesterolaemic effects, prevention of inflammatory bowel disease and effect on *Helicobacter pylori* eradication. Currently, most of probiotic strains used commercially were initially isolated from human feces whether adults or children, as well as dairy products. However, recently, it has been shown that non-fermented dairy foods such as vegetable foods are an excellent source of bacteria to find new probiotic strains with a significant probiotic potential. In this review, the history of probiotics, their mechanisms of action and the beneficial health effects are presented.

Keywords: *Bacillus*, Beneficial health effects, *Bifidobacterium*, *Enterococcus*, Immunomodulation, *Lactobacillus*, Mechanisms of action, Metabolic effects, Probiotics, Safety assessment, Technological and functional criteria, *Sacharomyces*.

1. INTRODUCTION

The term probiotic comes from the Greek 'pro bios' which means 'for life'. The interest for probiotics began with the history of man; fermented milk and cheese were well-known to the Greeks and Romans, who suggested their consumption, especially for convalescents and children. Introduction of the concept is generally

[#] **Corresponding author Antonio Gálvez:** University of Jaén, Department of Health Sciences, 23071-Jaen, Spain. E-mail: agalvez@ujaen.es

Manuela Oliveira and Isa Serrano (Eds)

attributed to Eli Metchnikoff, who in 1907 observed that Bulgarian peasant populations lived to very old ages; they had average life spans that stretched longer than wealthier European populations [1]. Their diets, Mechnikov annotated, were rich in yogurt and other fermented milk products [2].

Eli Metchnikoff related this to the 'Bulgarian *bacillus*' which he later demonstrated how healthy bacteria in yogurt helped digestion, improved the immune system and counteract the putrefactive effects of gastrointestinal (GI) metabolism [3]. His work at the Pasteur Institute in Paris reinforced his theories about the beneficial qualities of the lactic acid bacteria (LAB) produced by fermentation. He called the strain *Lactobacillus bulgaricus*. Mechnikov and his colleagues began drinking sour milk to populate their own gut flora, in this way introducing the modern probiotic [4].

The scientific rationale for the health benefit of lactic acid bacteria was given in his book "The prolongation of life" published in 1907 [5]. He suggested that the bacteria implicated in yoghurt fermentation, *Lactobacillus bulgaricus* and *Streptococcus thermophilus*, remove the putrefactive-type fermentations of the intestinal flora and that consumption of these yoghurts played a role in maintaining health [4]. Mechnikov in his book affirmed that some of the microorganisms present in the large intestine were a source of toxicants, that contributed to disease and aging. He found that "The dependence of the intestinal microbes on the food makes it possible to adopt measures to modify the flora in our bodies and to replace the prejudicial microbes by useful microbes" [5].

At the same time, another important discovery was made at the Pasteur Institute by Henri Tissier who demonstrated that children with diarrhea had a low number of bacteria characterized by a peculiar, Y shaped morphology. These "bifid" bacteria were, on the contrary, abundant in healthy children [6]. He found that these bacteria could be administered to patients with diarrhea to help restore a healthy gut flora. The works of Metchnikoff [1] and Tissier [6] were the first to make scientific suggestions concerning the probiotic use of bacteria.

According to Vergin [7], the term 'Probiotika' was originally coined by the physician and dietician Werner Kollath shortly after the Second World War, meaning 'food ingredients with health-promoting characteristics beyond their nutritional value' [8]. Vergin [7] suggested that the microbial imbalance in the body caused by antibiotic treatment could have been re-established by a probiotic rich diet; a suggestion cited by many as the first reference to probiotics as they are defined at the present time. Similarly, Kolb [9] recognized prejudicial effects of antibiotic treatment and proposed the prevention by probiotics [2].

Later and independently, Lilly and Stillwell, introduced the term 'probiotic' into the international literature in 1965, saying that a probiotic is a microbial substance secreted by one microorganism which stimulate the growth of another, thus being the exact opposite of an antibiotic [10]. Nevertheless, this definition did not gain wide acceptance and failed to survive. Since that time, the meaning of the term 'probiotic' has changed considerably. In 1974, Parker talks about a food supplement for livestock and improved name of probiotics as "organisms and substances which contribute to intestinal microbial balance" [11].

Some other definitions followed, with Fuller [12] being the first pointing out the microbial nature of probiotics by redefining the word "probiotic" as "a live microbial feed supplement that beneficially affects the host animal by improving its intestinal balance". Havenaar and coworkers [13] defined probiotics as 'mono- or mixed cultures of live microorganisms which, when applied to animal or man, beneficially affect the host by improving the property of the indigenous flora'. Guarner and Schaafsma [14] gave a more recent definition as "live microo-rganisms, which when consumed in adequate amounts, confer a health effect on the host".

The report of the joint Food and Agriculture Organization (FAO) and World Health Organization (WHO) redefined probiotics as: ''live microorganisms that, when administered in adequate amounts, confer a health benefit on the host'' (FAO/WHO) [15] (Table **1**). In 2002, the United Nations FAO/WHO Working Group generated new guidelines for the development and evaluation of probiotics found in foods (Fig. **1**).

2. SELECTION OF PROBIOTICS

The initial screening and selection of probiotics includes testing of the following important criteria:

- Safety assessment, including: strain identification and typing, virulence and pathogenic properties, antimicrobial resistance, metabolic activities associated with production of toxic substances, platelet-aggregating activity and mucus degradation activity.
- Technological criteria, including: strain authenticity and genetic stability, good sensory properties, phage-resistance, desired viability during processing and storage, and large-scale production.
- Functional criteria, including: Acid and bile stability (tolerance to gastric acid and juices, bile tolerance), adhesion stability (adhesion to mucosal surface), Viability and properties during processing and storage and validated and documented health effects.
- Desirable physiological criteria, including: antipathogenic activity (antagonistic activity towards gastrointestinal pathogens), immunomodulatory activity, inflammatory bowel disease prevention, cholesterol metabolism, lactose metabolism and antimutagenic and anticarcinogenic properties.

Table 1. Some of the definitions of probiotics cited over the years.

Definition	**Source**
Live microorganisms that, when administered in adequate amounts confer a health benefit on the host	[14, 15]
Microbial cell preparations or components of microbial cells that have a beneficial effect on the health and well-being of the host.	[17]
Probiotic is a microbial substance that is produced by one microorganism and has a growth-promoting effect on another microorganism.	[10]
Organisms and substances which contribute to intestinal microbial balance.	[11]
A live microbial feed supplement which beneficially affects the host animal by improving its intestinal microbial balance'.	[12]
Mono- or mixed cultures of live microorganisms which, when applied to animal or man, beneficially affect the host by improving the property of the indigenous flora.	[13]
Food ingredients with health-promoting characteristics beyond their nutritional value.	[16]
Probiotics are opposite of antibiotics.	[7]

(Table 1) contd.....

Definition	**Source**
Deleterious effects of antibiotics can be prevented by probiotic therapy.	[9]

Currently, most of probiotic strains used commercially were initially isolated from human feces whether adults or children, as well as dairy products. However, recently, it has been shown that non-fermented dairy foods such as vegetable foods are an excellent source of bacteria to find new probiotic strains with a significant probiotic potential [18 - 20].

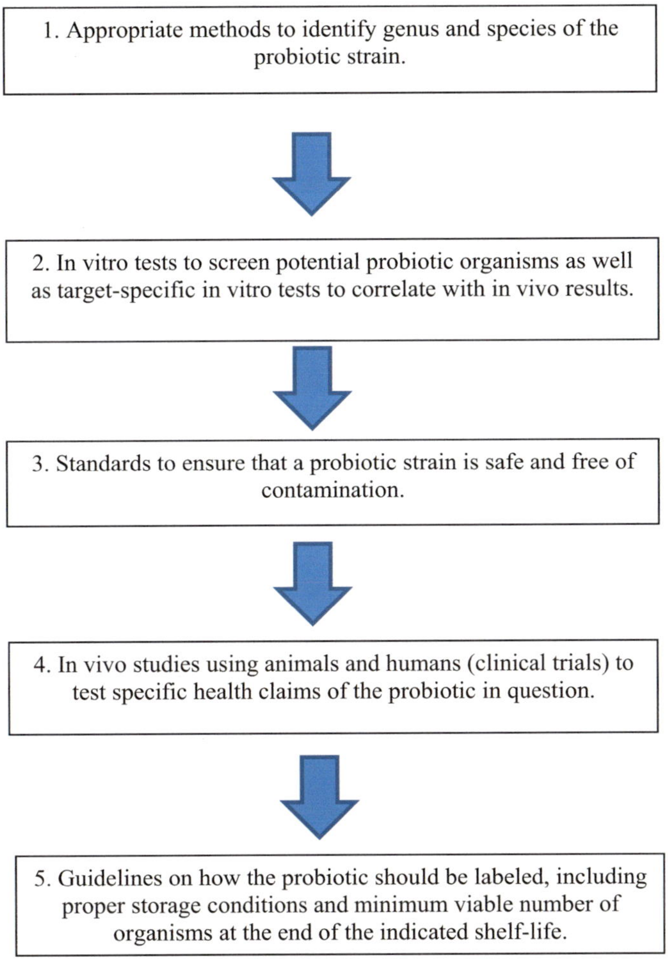

Fig. (1). Scheme of the guidelines for the evaluation of probiotics for food use (FAO/WHO).

In this sense, Abriouel and coworkers [20] found that *Lactobacillus pentosus* MP-10 strain isolated from brines of naturally fermented Aloreña green table olives, exhibits several probiotic and nutritional properties [20]. These results open a new perspective to investigate new strains with probiotic potential.

3. COMMERCIALLY IMPORTANT PROBIOTICS

Most of probiotic strains belong to the genera *Lactobacillus* and *Bifidobacterium*, nevertheless, other bacteria and some yeast may have probiotic properties (Table **2**). Moro (1900) was the first researcher to isolate facultative anaerobic straight rods from the faeces of breast-fed infants, which he typified as *Bacillus acidophilus*, a generic name for intestinal lactobacilli. Lactobacilli are usually described as Gram-positive, non-spore-forming and non-flagellated rods or cocobacilli. They are either aerotolerant or anaerobic and strictly fermentative [21].

Glucose is fermented predominantly to lactic acid in the homofermentative case, or equimolar amounts of lactic acid, CO_2 and ethanol (and or acetic acid) in the heterofermentative counterpart [22, 23]. Lactobacilli have complex growth requirements (low oxygen tension, protein and its breakdown products, vitamins) and growth may occur in a temperature range between 30 and 45°C.

Table 2. Microorganisms considered as probiotics.

Lactobacillus	*Bifidobacterium species*
L. acidophilus	*B. adolescentis*
L. bulgaricus	*B. animalis sp. lactis*
L. casei	*B. bifidum*
L. delbruekii sp. bulgaricus	*B. breve*
L. fermentum L. helveticus	*B. essencis*
L. lactis	*B. infantis*
L. paracasei	*B. lactis*
L. salivarus	*B. longum*
L. crispatus	***Other lactic acid bacteria***
L. gallinarum	
L. gasseri	*Enterococcus faecium*
L. johnsonii	*Lactococcus lactis*
L. plantarum	*Leuconostoc mesenteroides*
L. reuteri	*Pediococcus sp.*
L. rhamnosus	*Sporolactobacillus*
	Streptococcus sp.

(Table 2) contd.....

Non lactic acid bacteria	Yeasts
Bacillus cereus *Escherichia coli* strain Nissle *Propionibacterium freudenreichii*	*Saccharomyces cerevisiae* *S. boulardii*

Lactobacilli are ubiquitous and normally harmless. In humans and animals they are found in the gastrointestinal and genital tracts and perform many beneficial functions. They are rarely associated with cases of gastrointestinal and extraintestinal infection, and strains employed technologically are regarded as non-pathogenic and safe microorganisms. Furthermore, they have the reputation of health promoters, especially in the human gastrointestinal and genitourinary tracts [24]. Their distribution is affected by some environmental factor, that include level of specific substrates, pH, oxygen availability, presence of secretions and bacterial interactions.

On the other hand, Bifidobacteria were first isolated and described in 1899–1900 by Tissier which he termed *Bacillus bifidus*. Bifidobacteria are generally characterized as a Gram-positive, rod-shaped, non-gas-producing, non-spore forming, catalase negative and anaerobic microorganisms. They have various shapes including short, curved rods, club-shaped rods and bifurcated Y-shaped rods. They are saccharolytic organisms that produce acetic and lactic acid without generation of CO_2. Bifidobacteria are microorganisms of paramount importance in the active and complex ecosystem of the intestinal tract of humans. The indigenous microflora of infants is dominated by bifidobacteria, which are established shortly after birth. Their proliferation is stimulated by the glycoprotein components of k-casein in human colostrum and, to a lesser extent, human milk. The number of bifidobacteria decreases with increasing age of an individual and eventually becomes the third most abundant genus (accounting for approx. 25 % of the total adult gut flora) after the genera *Bacteroides* and *Eubacterium* [25].

Besides the probiotics strains belonging to *Lactobacillus* and *Bifidobacterium* genera, some other species of bacteria and fungi such as *Bacillus* [26], *Enterococcus* [27], and *Sacharomyces* [28] have also been used as probiotics. *Enterococcus* strains have been used as supplement for the food and feed such as poultry and swine to replace the use of sub-lethal antibiotics in the feeds. Many studies have been conducted to evaluate the effect of probiotic strains of this genus (mainly *Enterococcus faecium*). Generally, human studies are scarcer when compared to animal applications. This fact is due to the often association of enterococci with the pathogenesis such as infections of the urogenital tract, bacteremia, neonatal sepsis and endocarditis [29 - 30]. It is capable of transporting the antibiotic resistance genes to produce β-hemolysin, gelatinase and aggregation substance that are undesirable phenotypes in probiotic strain [29 - 30].

The scientific interest in *Bacillus* species as probiotics though, has only occurred in the last 15 years [31 - 33]. Of the species that have been most extensively examined these are *Bacillus subtilis*, *Bacillus clausii and Bacillus coagulans*. *B. subtilis* has been studied physiologically and genetically, and is associated with a Japanese product known as natto. The consumption of this fermented food can lead to stimulation of the immune system and reduction of blood coagulation by fibrinolysis [34, 35]. *B. coagulans* has been used as adjunct therapy for relieving rheumatoid arthritis [36]. The secretion of antimicrobials such as coagulin, amicoumacin and subtilisin is also verified in *Bacillus*.

The proposed mechanisms for probiotic effects of the *Bacillus* spores are based on immunomodulation, which occurs through the stimulation of the gut-associated lymphoid tissue by production of cytokines, competitive exclusion of gastrointestinal pathogens (*e.g.* competition for adhesion sites) and secretion of antimicrobial substances [37]. Spores being heat-stable have a number of advantages over other non-spore formers such as *Lactobacillus* spp., namely, that the product can be stored at room temperature in a desiccated form without any deleterious effect on viability. A second advantage is that the spore is capable of surviving the low pH of the gastric barrier [38, 39] which is not the case for all species of *Lactobacillus* [40] so in principle a specified dose of spores can be stored indefinitely without refrigeration and the entire dose of ingested bacteria will reach the small intestine intact.

However, genera such as *Enterococcus* and *Bacillus* despite being labeled as probiotics must be carefully considered since concerns still remain regarding the safe use of these organisms as probiotics [41].

The nonpathogenic yeast *Saccharomyces cerevisiae* and *S. cerevisiae* var.

boulardii had been prescribed in the past 30 years for prophylaxis and treatment of diarrheal diseases caused by bacteria (Table **3**). Some studies have shown that it is effective in preventing and treating diarrhea associated with the use of antibiotics and traveler's diarrhea. It has also been reported to prevent the reoccurrence of *Clostridium difficile*, to treat acne, and to reduce side effects of treatment for *Helicobacter pylori* [42].

Table 3. The efficacy of *Saccharomyces boulardii* as a probiotic.

Properties of *Saccharomyces boulardii*	**Sources**
Survives transit through the GI tract (most commonly the colon).	[43]
Its temperature optimum is 37 C, both *in vitro* and *in vivo*: unique advantage of being one of the few yeasts that do best at human body temperatures.	[44, 45]
Inhibits the growth of a number of microbial pathogens.	[42]
In lyophilized form, *S. boulardii* survives gastric acid and bile.	[44, 45]
When given orally, achieves steady-state concentrations within three days and is cleared within 3–5 days after it is discontinued.	[42]

4. MECHANISMS OF ACTION OF PROBIOTICS

Several mechanisms of probiotic action have been described, the most common may be classified in two modes of action (Fig. **2**).

- Direct effect on other microorganisms (commensal and/or pathogenic microorganisms).
- Immunomodulation: probiotics might be capable to modulate the host's defences (including the innate as well as the acquired immune system).

4.1. Immunomodulation

The mechanisms by which epithelial cells are making the difference between probiotic and pathogenic microorganisms appear to be different. Pathogenic bacteria induce a pro-inflammatory response in epithelial cells by activating transcription factor NF-κB. Compared with these bacteria, non-pathogenic species may alleviate to the side of pro-inflammatory response by blocking this factor [46].

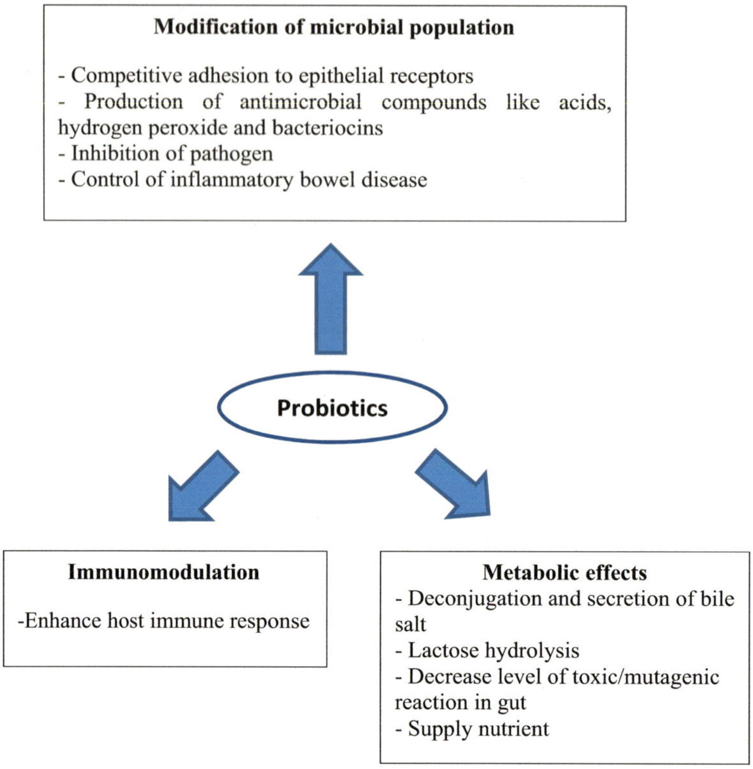

Fig. (2). Mechanism of action of probiotics.

This concept was demonstrated by Lammers and coworkers [47] and Otte and Podolsky [48] who showed that probiotic bacteria in the VSL#3 combination did not induce IL-8 secretion by epithelial cells compared with intestinal pathogens such as enteropathogenic *Escherichia coli*, *Salmonella* Dublin, *Shigella dysenteriae*, and *Listeria monocytogenes*, all of which did induce secretion of IL-8 [47, 48]. Similarly Zhang and coworkers [49] demonstrated that *Lactobacillus rhamnosus* GG reduces the amount of IL-8. However, the probiotic bacteria *E. coli* Nissle 1917 induced IL-8 secretion by intestinal epithelial cell lines in a dose-dependent manner, suggesting that the ability to prevent secretion of IL-8 from epithelial cells is not a feature of all probiotic bacteria [50]. During the last years several studies have shown the influence of probiotics on different immune cells such as dendritic cells, monocytes and macrophages, and on lymphocytes and T-cell redistribution (Fig. **3**) [50].

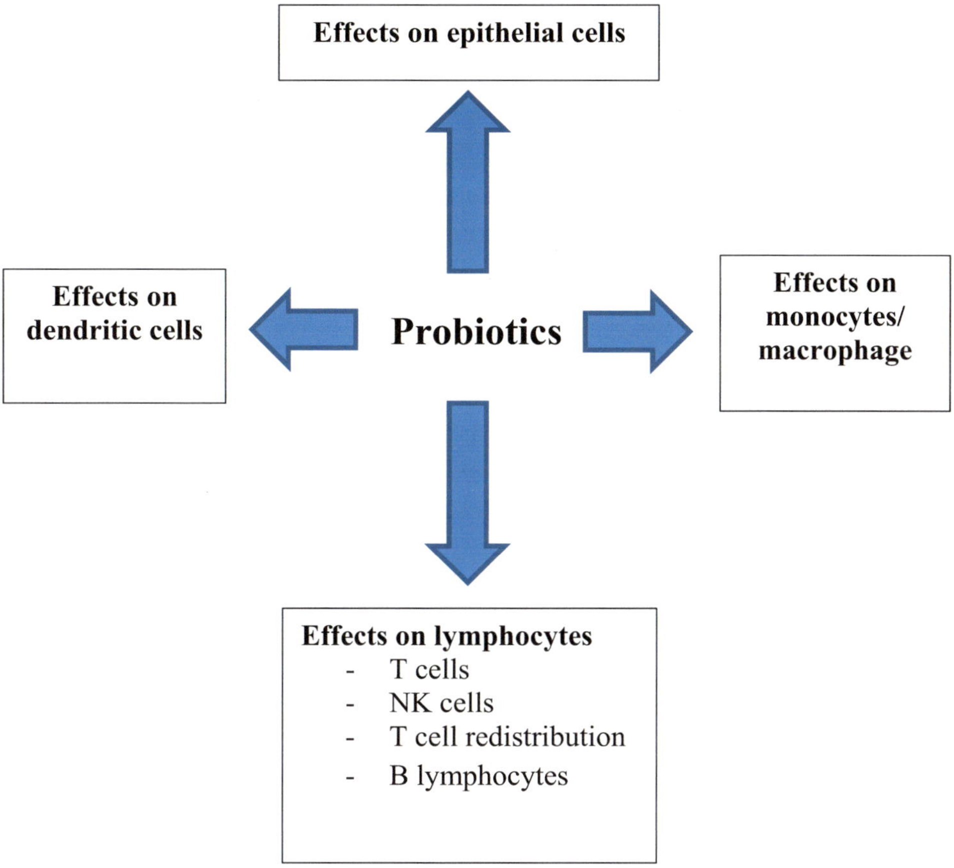

Fig. (3). The effects of probiotics on numerous cell types involved in the innate and adaptive immune responses.

4.2. Direct Effects on Other Microorganisms

Probiotics have a direct effect on other microorganisms, pathogenic and/or commensal ones. This principle is in many cases of importance for the prevention and therapy of infections and re-establishment, of the microbial equilibrium in the gut. Probiotic bacteria can antagonize pathogenic bacteria by decreasing luminal pH, inhibiting bacterial invasion, blocking bacterial adhesion to epithelial cells or producing bacteriocins.

Bacteriocin production has been a valuable feature in the selection of a probiotic strain, although, that not many research have definitely showed the impact of bacteriocin production on the capacity of a strain to compete within the GI tract and positively influence the health of the host [51]. Gillor and coworkers [52] showed that colicin produced by *E. coli* was capable to persist in the intestine of streptomycin-treated mice for an prolonged period of time relative to their non-colicin-producing counterparts. Hillman and coworkers [53] demonstrated a good correlation between the capacity of a *Streptococcus mutans* strain to colonize the oral cavity and the production of the bacteriocin mutacin 1140. The study by Bhardwaj and coworkers [54] demonstrated that the bacteriocin- producing strain *Enterococcus faecium* KH24 considerably affected the fecal microbiota of mice.

In the case of *Bifidobacterium,* Lee and coworkers [55] demonstrated that *Bifidobacterium longum* DJO10A had a significantly greater ability to compete against the intestinal isolates *Clostridium difficile* DJOcd1 and *Escherichia coli* DJOec1 than did the nonproducing isogenic variant *B. longum* DJO10-JH1. In this sense the results found by Rea and coworkers [56] showed that Lacticin 3147, a broad-spectrum bacteriocin produced by *Lactococcus lactis* subsp. *lactis*, inhibits a range of genetically distinct *C. difficile* isolates from healthy subjects and patients with inflammatory bowel disease (IBD).

Regarding genus *Lactobacillus,* Flynn and coworkers [57] showed that the probiotic *L. salivarius* subsp. *salivarius* UCC118 produces a peptide that inhibits a broad range of pathogens such as *Bacillus*, *Staphylococcus*, *Enterococcus*, *Listeria*, and *Salmonella* species. Gotteland and coworkers [58] showed the antimicrobial effect of *Lactobacillus* species on *Helicobacter pylori* infection of gastric mucosa, achieved by the release of bacteriocins and the ability to decrease adherence of this pathogen to epithelial cells.

Although extensive progress has been made with respect to our understanding of bacteriocin we thought that additional research is required to gain a full understanding of the factors which control bacteriocin production in the GI tract [59].

5. BENEFICIAL HEALTH EFFECTS OF PROBIOTICS

The beneficial effects of probiotic foods on human health and nutrition are increasingly recognized by health professionals. This review will be focused on the following beneficial health effects of probiotics.

- Alleviation of lactose intolerance
- Effect on *Helicobacter pylori* eradication
- Anti-cancer effect
- Allergy and atopic dermatitis
- Cholesterol-Lowering Effects
- Inflammatory diseases and bowel syndromes

Table 4. Positive effects of probiotics on human health.

Microorganisms	Strains	Effects on human health
Bifidobacteria	*Bifidobacterium animalis* DN-173 010 (marketed as: *Bifidus Regularis*)	Gut health and digestion
	Bifidobacterium infantis 35624	Helps irritable bowel syndrome
	Bifidobacterium lactis Bb-12	Immune system and gut health
	Bifidobacterium bifidum	Immune system and gut health
	Bifidobacterium breve	Immune system and gut health
Lactobacilli	*Lactobacillus casei Shirota*	Immune system and gut health
	Lactobacillus casei DN-114 001 (marketed as: *L. casei Immunitas*)	Immune system, lessens duration of colds and flues in older people
	Lactobacillus rhamnosus GR-11	Helps vaginal infections
	Lactobacillus reuteri 55730	Helps antibiotic associated diarrhea
Yeast	*Saccharomyces boulardii*	Helps antibiotic associated diarrhea

It is essential to take into account that health benefits provided by probiotics are strain specific, and not genus or species specific. Accordingly, no probiotic strain will provide all proposed benefits, not even strains of the same species, and not all strains of the same species will be effective against defined health conditions [60]. A summary of the positive effects of probiotics on human health is provided in Table **4**.

5.1. Alleviation of Lactose Intolerance

Lactose intolerance is the most common type of carbohydrate malabsorption. It is associated with the inability to digest lactose into its constituents, glucose and galactose, due to low levels of lactase enzyme activity [61]. It is generally accepted that fermented milk products such as yogurt can efficiently improve lactose digestion in lactose malabsorbers and therefore that they are well tolerated by most lactose-intolerant subjects [61]. These beneficial effects are due to the presence of lactase producing bacteria in the yogurt, especially *Lactobacillus bulgaricus* and *Streptococcus thermophilus*, contributed to the digestion and absorption of lactose [61]. It was also found that the presence of *Lactobacillus acidophilus* alleviate lactose intolerance through their ability to produce lactase enzyme [61].

Finally, in another study it was found that consumption of milk containing *Bifidobacterium longum* resulted in significantly less hydrogen production and flatulence as compared to the consumption of control pasteurized milk [61]. Vonk and coworkers [62] showed that a mix of probiotics in yogurt together with *B. longum* capsules could increase the β-galactosidase activity in faeces and alleviate the complaints of lactose intolerance.

5.2. Effect on *Helicobacter Pylori* Eradication

There are several reports about the action of the probiotics against *Helicobacter pylori* [63, 64], a Gram negative pathogen responsible for chronic gastritis and peptic ulcer and a risk factor for gastric malignancies. Probiotics are believed to have a role in eradicating and possibly preventing *H. pylori* infection as an adjunctive treatment. Literature has shown that lactobacilli or their cell-free cultures inhibit *H. pylori*, prevent its adhesion to mammalian epithelial cells and prevent interleukin-8 release.

The beneficial effects of fermented milk containing *Lactobacillus sp.* on *H. pylori* infection have also been reported [65, 66]. However, the administration of probiotics alone does not eradicate *H. pylori*, and after stopping ingestion, the suppressing effect on *H. pylori* was lost. Sheu and coworkers [67] reported that pre-treatment with *Lactobacillius* and *Bifidobacterium* containing yogurt improved the efficacy of quadruple therapy and decreased bacterial load [67].

Similar results were *obtained* by Hamilton-Miller and coworkers [68] who reported that yogurt-containing *Lactobacillus gasseri* (OLL2716) had a suppressive effect on *H. pylori* infection.

Dajani and coworkers [69] showed that the probiotic *B. infantis* appeared to be a sound alternative to triple therapy, when added as an adjuvant. These authors observed that *B. infantis* led to better results of eradication for people who had been previously treated with clarithromycin, that is, reduced the negative effect of clarithromycin resistance. *B. infantis* also improved significantly the clinical symptoms and the incidence of antibiotic induced diarrhea with only few and milder incidents of diarrhea reported [69].

5.3. Anti-carcinogenic Effect

Probiotics have a promising future in the prevention of several cancers. The most promising evidence is with regards to colon cancer [70]. The anti-carcinogenic effect may be attributable to a combination of mechanisms such as the induction of pro-inflammatory, anti-inflammatory or secretory responses that might inhibit carcinogenesis.

Several studies have shown that preparations containing LAB inhibit the growth of tumor cells in experimental animals or indirectly lower carcinogenicity by decreasing bacterial enzymes that activate carcinogenesis [71]. Lactobacilli and bifidobacteria have shown the ability to decrease the genotoxic activity of certain chemical compounds [72] and increase in antimutagenic activity during the growth in selected media [73].

Chiu and coworkers [74] described bacterial soluble factors *(LcrS)* secreted by *Lactobacillus casei* and *L. rhamnosus* which induced apoptosis of human monocytic leukemia-cell line (THP-1). Studies in animal models also suggested that crescent natural killer cell activity by probiotic consumption may have positive effects on delayed tumor development.

Studies by Lyer and coworkers [75] have demonstrated that some probiotic strains have influenced hematological cancers. *Lactobacillus reuteri* enhanced TNF-induced apoptosis in human chronic myeloid leukemia-derived cells by modulation of NF-κB and MAPK signaling and reduced proteins that mediated cell proliferation (cyclin D1 and COX-2) or inhibited apoptosis (Bcl-2, Bcl-xL). Short-chain fatty acids produced by *L. acidophilus* and bifidobacteria were also reported to inhibit the generation of carcinogenic products by reducing enzyme activities. When incubated *in vitro* with 4-nitroquinoline-1-oxide (4NQO), some probiotic strains inhibited the genotoxic activity of 4NQO. *L. casei* was most effective, followed by *Lactobacillus plantarum* and *L. rhamnosus* [76].

Lactobacillus and *Bifidobacteria* strains and *E. coli* strain Nissle 1917 have shown anti-mutagenic activities *in vitro*, probably due to their capacity to metabolize and inactivate mutagenic compounds [77]. Another study reported that exopolys accharides (EPS) of *Lactobacillus acidophilus* were involved in colon cancer cell death *via* autophagy [78].

Cytoplasmic fractions of *L. casei* YIT9029 and *B. longum* HY8001 showed the ability to suppress the proliferation of tumor cells when administrated orally to mice as dietary supplement [79]. Baldwin and coworkers [80] showed that *L. casei and L. acidophilus* were capable to increment the apoptosis-induction capacity of 5-fluorouracil in colorectal carcinoma cell line LS513, indicating that these probiotics may be used as adjuvants in anticancer chemotherapy. Altonsy and coworkers [81] described induction of mitochondrial pathway of apoptosis in human colonic carcinoma cells (Caco-2) incubated with probiotics such as *Bifidobacterium lactis or L. rhamnosus* or commensal bacteria *Atopobium minutum* and *E. coli.* However both studies did not determine whether examined probiotics had similar abilities to affect apoptosis in human normal colon cells [82].

In conclusion, the anticancer activity through induction of apoptosis of cancer cells seems to be promising approach for use of some probiotic strains as a support therapy or cancer prevention.

5.4. Cholesterol-lowering Effects

Since Mann and Spoerry first reported the hypocholesterolemic activity of fermented milk in the Maasai tribe of Kenya in 1974 [83], several scientists have used animal models or humans to evaluate the effects of probiotics on serum cholesterol levels, whose benefits have been emphasized over the years [83]. Several mechanisms have been hypothesized (Table **5**).

5.5. Allergy and Atopic Dermatitis

Several studies have suggested that probiotics could protect against allergies. Their effect in the diminution of allergic diseases could be tied to the hygiene hypothesis, which indicate that an absence of exposure to microbes early in life can affect the development of the immune system and increment susceptibility to allergies. Allergic diseases are associated with an imbalance in T helper (TH1/TH2) cytokine activation [91]. An insufficient stimulation of Th1 cells cannot offset the expansion of Th2 cells and results in a predisposition to allergy [92].

Table 5. Mechanisms of the cholesterol-lowering effects of probiotics.

Mechanisms of the cholesterol-lowering effects of probiotics	Source
Enzymatic deconjugation of bile acids by bile-salt hydrolase of probiotics.	[84]
Assimilation of cholesterol by probiotics.	[85]
Co-precipitation of cholesterol with deconjugated bile.	[86]
Cholesterol binding to cell walls of probiotics.	[87]
Incorporation of cholesterol into the cellular membranes of probiotics during growth.	[88]
Conversion of cholesterol into coprostanol.	[89]
Production of short-chain fatty acids upon fermentation by probiotics in the presence of prebiotics.	[90]

Many studies have demonstrated that atopic children have a different intestinal microflora from that of healthy children, with lower levels of bifidobacteria and higher levels of Clostridia. Moreover, other research studies have also demonstrated that early colonization with potentially pathogenic bacteria such as *Staphylococcus aureus* and *Clostridium difficile* is more likely to occur in children who go on to develop allergy. However, lactobacilli and bifidobacteria are present more commonly in the composition of the intestinal flora of non-allergic children. The enhanced presence of these probiotic bacteria in the intestinal microbiota seems to correlate with protection against atopy [93, 94].

Gueimonde and coworkers [95] demonstrated that early consumption of probiotic preparations containing *Lactobacillus* GG may reduce prevalence of atopic eczema later in life. *L. rhamnosus* HN001 was reported effective against eczema in the first two years of life, persisting to age 4 years, while *B. animalis* subsp.

Lactis HN019 had no effect [96]. Isolauri and coworkers [97] evaluated the effect of the hydrolysed whey formulas supplemented with probiotics *B. lactis* Bb12 and *L. rhamnosus* GG in 27 breast-fed infants that suffered from atopic eczema. These authors found that clinical signs and symptoms of atopic eczema were diminished after two months in both groups [23]. However, according to Vandenplas and coworkers [96] the data on probiotics and allergy need further clarification, because data are some-how contradictory.

5.6. Inflammatory Diseases and Bowel Syndromes

Inflammatory diseases and bowel syndromes comprises a spectrum of disorders characterized by inflammation, ulceration and abnormal narrowing of the gastrointestinal tract resulting in abdominal pain, diarrhea and gastrointestinal bleeding [98, 99]. Several studies demonstrated that the intestinal microflora likely plays a critical role in inflammatory conditions in the gut and potentially probiotics could remediate such conditions through modulation of the microflora. [2]. Probiotics are used in the treatment and management of inflammatory bowel disease (IBD), which includes a variety of GI diseases, such as Crohn's disease [100], pouchitis [101], and ulcerative colitis [102]. Nanda and coworkers [103] evaluated *in vivo* assays with probiotic bacteria the effects of an acute or chronic gut inflammation using dextran sulphate sodium (DSS), which induces colitis in mice, presenting positive results. The American Society for Nutrition released guidelines that support the use of probiotics for ulcerative colitis and pouchitis patients based on research that showed effectiveness with *E coli* Nissle 1917 and VSL#3 that contains eight strains of bacteria [104]. Kelesidis and Pothoulakis [42] showed the positive effect of *S. boulardii* in patients with IBD. In Table **6** we can see the list of probiotics that have been studied in humans for their potential clinical relevancy in ulcerative colitis, Crohn's disease, and pouchitis.

Table 6. List of probiotics with potential clinical relevancy in ulcerative colitis, Crohn's disease, and pouchitis.

GI diseases	Probiotics
Ulcerative colitis	*Escherichia coli* Nissle 1917, *Bifidobacterium breve*, *Bifidobacterium longum*, *Bifidobacterium infantis*, *Lactobacillus acidophilus*, *Lactobacillus plantarum*, *Lactobacillus casei*, *Lactobacillus bulgaricus*, *Streptococcus thermophilus*; *Lactobacillus boulardi*, *Lactobacillus rhamnosus* GG.

(Table 6) contd.....

GI diseases	Probiotics
Crohn's disease	*Lactobacillus boulardi, Lactobacillus rhamnosus* GG, *Lactobacillus johnsonii, Bifidobacterium breve, Bifidobacterium longum, Bifidobacterium infantis, Lactobacillus acidophilus, Lactobacillus plantarum, Lactobacillus casei, Lactobacillus bulgaricus, Streptococcus thermophilus*; *Lactobacillus boulardi, Saccharomyces boulardii.*
Pouchitis	*Bifidobacterium breve, Bifidobacterium longum, Bifidobacterium infantis, Lactobacillus acidophilus, Lactobacillus plantarum, Lactobacillus casei, Lactobacillus bulgaricus, Streptococcus thermophilus*; *Lactobacillus boulardi, Lactobacillus* GG.

CONCLUSION

In the last years, probiotics have been widely studied showing beneficial effects to human and animal health. Also, it was largely argued that health benefits provided by probiotics are strain specific, and not genus or species specific. Accordingly, no probiotic strain will provide all proposed benefits, not even strains of the same species, and not all strains of the same species will be effective against defined healthy conditions. Currently, most of probiotic strains commercially used were initially isolated from human feces whether adults or children, as well as dairy products. However, recently, it has been shown that non-fermented dairy foods such as vegetable foods are an excellent source of bacteria to find new probiotic strains with a significant probiotic potential. This may be the future in the research of new probiotic strains. Also, the application of molecular methods in probiotic studies is revolutionizing our understanding of them. In this sense, genomics is one of the research areas in which they are placing greater expectations for the future.

CONFLICT OF INTEREST

The authors confirm that this chapter contents have no conflict of interest.

ACKNOWLEDGEMENTS

Declared none.

REFERENCES

[1] Metchnikoff E. Optimistic studies. New York: Putman's Sons 1908; pp. 161-83.

[2] Azizpour K, Bahrambeygi S, Mahmoodpour S, *et al.* History and basic of probiotics. Res J Biological Sci 2009; 4: 409-26.

[3] Kingsley CA, Gregor R. Probiotics: 100 years (1907-2007) after Elie Metchnikoff's observation. In: Méndez-Vilas A, Ed. Communicating Current Research and Educational Topics and Trends in Applied Microbiology. (2007 ed.) Spain: Formatex.org 2008; pp. 466-74.

[4] Soccol CR, Vandenberghe LP, Spie MR, *et al.* The Potential of Probiotics: A Review. Food Technol Biotechnol 2010; 48: 413-34.

[5] Metchnikoff E. The Prolongation of Life: Optimistic Studies. New York, USA: Springer Publishing Company 2004.

[6] Tissier H. The treatment of intestinal infections by the method of transformation of bacterial intestinal flora. C R Soc Biol 1906; 60: 359-61.

[7] Vergin F. Anti- und Probiotika. Hippokrates 1954; 25(4): 116-9.
[PMID: 13174124]

[8] Sonnenborn U, Schulze J. The non-pathogenic *Escherichia coli* strain Nissle 1917 – features of a versatile probiotic. Microb Ecol Health Dis 2009; 21: 122-58.
[http://dx.doi.org/10.3109/08910600903444267]

[9] Kolb H. Die Behandlung acuter infekte unter dem gesichtswinkel der prophylaxes chronischer leiden. Uber die behandlung mit physiologischen bakterien. Microecol Ther 1955; 1: 15-9.

[10] Lilly DM, Stillwell RH. Probiotics: growth-promoting factors produced by microorganisms. Science 1965; 147(3659): 747-8.
[http://dx.doi.org/10.1126/science.147.3659.747] [PMID: 14242024]

[11] Parker RB. Probiotics, the other half of the antibiotic story. Anim Nutr Health 1974; 29: 4-8.

[12] Fuller R. Probiotics in man and animals. J Appl Bacteriol 1989; 66(5): 365-78.
[http://dx.doi.org/10.1111/j.1365-2672.1989.tb05105.x] [PMID: 2666378]

[13] Havenaar R, Huisin't Veld JH. Probiotics: A general view. The Lactic Acid Bacteria, Vol 1: The Lactic Acid Bacteria in Health and Diseas. New York, NY: Chapman & Hall 1992.
[http://dx.doi.org/10.1007/978-1-4615-3522-5_6]

[14] Guarner F, Schaafsma GJ. Probiotics. Int J Food Microbiol 1998; 39(3): 237-8.
[http://dx.doi.org/10.1016/S0168-1605(97)00136-0] [PMID: 9553803]

[15] Guidelines for the evaluation of probiotics in food, Report of a Joint FAO/WHO Working Group on Drafting Guidelines for the Evaluation of Probiotics in Food, London, Ontario, Canada 2002. (http://ftp.fao.org/es/esn/food/wgreport2.pdf).

[16] Kollath W. [Nutrition and the tooth system; general review with special reference to vitamins]. Dtsch Zahnarztl Z 1953; 8(11): 7-16.
[PMID: 13068115]

[17] Salminen S, Ouwehand A, Benno Y, Lee YK. Probiotics: How should they be defined? Trends Food Sci Technol 1999; 10: 107-10.
[http://dx.doi.org/10.1016/S0924-2244(99)00027-8]

[18] Abriouel H, Benomar N, Pulido RP, Cañamero MM, Gálvez A. Annotated genome sequence of Lactobacillus pentosus MP-10, which has probiotic potential, from naturally fermented Aloreña green table olives. J Bacteriol 2011; 193(17): 4559-60. a

[http://dx.doi.org/10.1128/JB.05171-11] [PMID: 21705590]

[19] Abriouel H, Benomar N, pérez pulido R, Martínez Cañamero M, Gálvez A. Mucosal adhesión properties of the probiotic *Lactobacillus pentosus* MP-10 isolated from naturally fermented Aloreña green table olives. III Workshop Probióticos, Prebióticos y Salud: Evidencia Científica. Castelldefels Barcelona 2011. b

[20] Abriouel H, Benomar N, Cobo A, *et al.* Characterization of lactic acid bacteria from naturally-fermented Manzanilla Aloreña green table olives. Food Microbiol 2012; 32(2): 308-16. [http://dx.doi.org/10.1016/j.fm.2012.07.006] [PMID: 22986194]

[21] Hammes WP, Vogel RF. The Genus *Lactobacillus'* in The Lactic Acid Bacteria In: Wood BJB, Holzapfel WH, Eds. The Genera of Lactic Acid Bacteria. Blackie Academic: London 1995; 2: pp. 19-54.

[22] Vasiljevic T, Shah NP. Probiotics –From Metchnikoff to bioactives. Int Dairy J 2008; 18: 714-28. [http://dx.doi.org/10.1016/j.idairyj.2008.03.004]

[23] Soccol CR, Porto de Souza L, Rigon M, Pedroni A, Tiemi C. The potential of probiotics: A review. Food Technol Biotechnol 2010; 48: 413-134.

[24] Salminen S, Isolauri E, Salminen E. Clinical uses of probiotics for stabilizing the gut mucosal barrier: successful strains and future challenges. Antonie van Leeuwenhoek 1996; 70(2-4): 347-58. [http://dx.doi.org/10.1007/BF00395941] [PMID: 8992950]

[25] Finegold SM, Sutter VL, Mathisen GE. Normal indigenous intestinal flora. In: Hentges DJ, Ed. Human intestinal microflora in health and Disease. New York, NY, USA: Academic Press 1983; pp. 3-31.
[http://dx.doi.org/10.1016/B978-0-12-341280-5.50007-0]

[26] Endres JR, Qureshi I, Farber T, *et al.* One-year chronic oral toxicity with combined reproduction toxicity study of a novel probiotic, *Bacillus coagulans*, as a food ingredient. Food Chem Toxicol 2011; 49(5): 1174-82.
[http://dx.doi.org/10.1016/j.fct.2011.02.012] [PMID: 21338652]

[27] Franz CM, Huch M, Abriouel H, Holzapfel W, Gálvez A. Enterococci as probiotics and their implications in food safety. Int J Food Microbiol 2011; 151(2): 125-40. [http://dx.doi.org/10.1016/j.ijfoodmicro.2011.08.014] [PMID: 21962867]

[28] Psomas E, Andrighetto C, Litopoulou-Tzanetaki E, Lombardi A, Tzanetakis N. Some probiotic properties of yeast isolates from infant faeces and Feta cheese. Int J Food Microbiol 2001; 69(1-2): 125-33.
[http://dx.doi.org/10.1016/S0168-1605(01)00580-3] [PMID: 11589551]

[29] Sánchez Valenzuela A, Benomar N, Abriouel H, *et al.* Virulence factors, antibiotic resistance and bacteriocins in enterococci from artisan foods of animal origin. Food Contr 2009; 20: 381-5. [http://dx.doi.org/10.1016/j.foodcont.2008.06.004]

[30] Valenzuela AS, Benomar N, Abriouel H, Cañamero MM, Gálvez A. Isolation and identification of *Enterococcus faecium* from seafoods: antimicrobial resistance and production of bacteriocin-like substances. Food Microbiol 2010; 27(7): 955-61.
[http://dx.doi.org/10.1016/j.fm.2010.05.033] [PMID: 20688238]

[31] Hong HA, Duc H, Cutting SM. The use of bacterial spore formers as probiotics. FEMS Microbiol Rev 2005; 29(4): 813-35.
[http://dx.doi.org/10.1016/j.femsre.2004.12.001] [PMID: 16102604]

[32] Mazza P. The use of *Bacillus subtilis* as an antidiarrhoeal microorganism. Boll Chim Farm 1994; 133(1): 3-18.
[PMID: 8166962]

[33] Sanders ME, Morelli L, Tompkins TA. Sporeformers as human probiotics: *Bacillus, Sporolactobacillus*, and *Brevibacillus*. Compr Rev Food Sci Food Saf 2003; 2: 101-10.
[http://dx.doi.org/10.1111/j.1541-4337.2003.tb00017.x]

[34] Sumi H, Yatagai C, Wada H, Yoshida E, Maruyama M. [Effect of *Bacillus* natto-fermented product (BIOZYME) on blood alcohol, aldehyde concentrations after whisky drinking in human volunteers, and acute toxicity of acetaldehyde in mice]. Arukoru Kenkyuto Yakubutsu Ison 1995; 30(2): 69-79.
[PMID: 7598662]

[35] Hosoi T, Kiuchi K. Production and probiotic effects of natto. In: Ricca E, Henriques AO, Cutting SM, Eds. Bacterial spore formers: probiotics and emerging applications. Wymondham, UK: Horizon Bioscience 2004; pp. 143-54.

[36] Mandel DR, Eichas K, Holmes J. *Bacillus coagulans*: a viable adjunct therapy for relieving symptoms of rheumatoid arthritis according to a randomized, controlled trial. BMC Complement Altern Med 2010; 10: 1.
[http://dx.doi.org/10.1186/1472-6882-10-1] [PMID: 20067641]

[37] Fuller R. Probiotics in human medicine. Gut 1991; 32(4): 439-42.
[http://dx.doi.org/10.1136/gut.32.4.439] [PMID: 1902810]

[38] Barbosa TM, Serra CR, La Ragione RM, Woodward MJ, Henriques AO. Screening for *bacillus* isolates in the broiler gastrointestinal tract. Appl Environ Microbiol 2005; 71(2): 968-78.
[http://dx.doi.org/10.1128/AEM.71.2.968-978.2005] [PMID: 15691955]

[39] Spinosa MR, Braccini T, Ricca E, *et al.* On the fate of ingested *Bacillus* spores. Res Microbiol 2000; 151(5): 361-8.
[http://dx.doi.org/10.1016/S0923-2508(00)00159-5] [PMID: 10919516]

[40] Tuohy KM, Pinart-Gilberga M, Jones M, Hoyles L, McCartney AL, Gibson GR. Survivability of a probiotic *Lactobacillus casei* in the gastrointestinal tract of healthy human volunteers and its impact on the faecal microflora. J Appl Microbiol 2007; 102(4): 1026-32.
[PMID: 17381746]

[41] Senok AC, Ismaeel AY, Botta GA. Probiotics: facts and myths. Clin Microbiol Infect 2005; 11(12): 958-66.
[http://dx.doi.org/10.1111/j.1469-0691.2005.01228.x] [PMID: 16307549]

[42] Kelesidis T, Pothoulakis C. Efficacy and safety of the probiotic *Saccharomyces boulardii* for the prevention and therapy of gastrointestinal disorders. Therap Adv Gastroenterol 2012; 5(2): 111-25.
[http://dx.doi.org/10.1177/1756283X11428502] [PMID: 22423260]

[43] Gorbach SL. Probiotics and gastrointestinal health. Am J Gastroenterol 2000; 95(1) (Suppl.): S2-4.
[http://dx.doi.org/10.1016/S0002-9270(99)00806-0] [PMID: 10634218]

[44] Graff S, Chaumeil JC, Boy P, Lai-Kuen R, Charrueau C. Formulations for protecting the probiotic *Saccharomyces boulardii* from degradation in acidic condition. Biol Pharm Bull 2008; 31(2): 266-72. a
[http://dx.doi.org/10.1248/bpb.31.266] [PMID: 18239285]

[45] Graff S, Chaumeil JC, Boy P, Lai-Kuen R, Charrueau C. Influence of pH conditions on the viability of *Saccharomyces boulardii* yeast. J Gen Appl Microbiol 2008; 54(4): 221-7. b
[http://dx.doi.org/10.2323/jgam.54.221] [PMID: 18802321]

[46] Neish AS, Gewirtz AT, Zeng H, *et al.* Prokaryotic regulation of epithelial responses by inhibition of IkappaB-alpha ubiquitination. Science 2000; 289(5484): 1560-3.
[http://dx.doi.org/10.1126/science.289.5484.1560] [PMID: 10968793]

[47] Lammers KM, Helwig U, Swennen E, *et al.* Effect of probiotic strains on interleukin 8 production by HT29/19A cells. Am J Gastroenterol 2002; 97(5): 1182-6.
[http://dx.doi.org/10.1111/j.1572-0241.2002.05693.x] [PMID: 12014725]

[48] Otte JM, Podolsky DK. Functional modulation of enterocytes by gram-positive and gram-negative microorganisms. Am J Physiol Gastrointest Liver Physiol 2004; 286(4): G613-26.
[http://dx.doi.org/10.1152/ajpgi.00341.2003] [PMID: 15010363]

[49] Zhang L, Li N, Caicedo R, Neu J. Alive and dead *Lactobacillus rhamnosus* GG decrease tumor necrosis factor-alpha-induced interleukin-8 production in Caco-2 cells. J Nutr 2005; 135(7): 1752-6.
[PMID: 15987860]

[50] Ng SC, Hart AL, Kamm MA, Stagg AJ, Knight SC. Mechanisms of action of probiotics: recent advances. Inflamm Bowel Dis 2009; 15(2): 300-10.
[http://dx.doi.org/10.1002/ibd.20602] [PMID: 18626975]

[51] Corr SC, Li Y, Riedel CU, O'Toole PW, Hill C, Gahan CG. Bacteriocin production as a mechanism for the antiinfective activity of *Lactobacillus salivarius* UCC118. Proc Natl Acad Sci USA 2007; 104(18): 7617-21.
[http://dx.doi.org/10.1073/pnas.0700440104] [PMID: 17456596]

[52] Gillor O, Giladi I, Riley MA. Persistence of colicinogenic *Escherichia coli* in the mouse gastrointestinal tract. BMC Microbiol 2009; 9: 165.
[http://dx.doi.org/10.1186/1471-2180-9-165] [PMID: 19674447]

[53] Hillman JD, Dzuback AL, Andrews SW. Colonization of the human oral cavity by a *Streptococcus mutans* mutant producing increased bacteriocin. J Dent Res 1987; 66(6): 1092-4.
[http://dx.doi.org/10.1177/00220345870660060101] [PMID: 3476580]

[54] Bhardwaj A, Gupta H, Kapila S, Kaur G, Vij S, Malik RK. Safety assessment and evaluation of probiotic potential of bacteriocinogenic *Enterococcus faecium* KH 24 strain under *in vitro* and *in vivo* conditions. Int J Food Microbiol 2010; 141(3): 156-64.
[http://dx.doi.org/10.1016/j.ijfoodmicro.2010.05.001] [PMID: 20570005]

[55] Lee JH, Karamychev VN, Kozyavkin SA, *et al.* Comparative genomic analysis of the gut bacterium *Bifidobacterium longum* reveals loci susceptible to deletion during pure culture growth. BMC Genomics 2008; 9: 247.
[http://dx.doi.org/10.1186/1471-2164-9-247] [PMID: 18505588]

[56] Rea MC, Clayton E, O'Connor PM, *et al.* Antimicrobial activity of lacticin 3,147 against clinical *Clostridium difficile* strains. J Med Microbiol 2007; 56(Pt 7): 940-6.
[http://dx.doi.org/10.1099/jmm.0.47085-0] [PMID: 17577060]

[57] Flynn S, van Sinderen D, Thornton GM, Holo H, Nes IF, Collins JK. Characterization of the genetic locus responsible for the production of ABP-118, a novel bacteriocin produced by the probiotic bacterium *Lactobacillus salivarius* subsp. *salivarius* UCC118. Microbiology 2002; 148(Pt 4): 973-84.
[PMID: 11932444]

[58] Gotteland M, Brunser O, Cruchet S. Systematic review: are probiotics useful in controlling gastric colonization by *Helicobacter pylori?* Aliment Pharmacol Ther 2006; 23(8): 1077-86.
[http://dx.doi.org/10.1111/j.1365-2036.2006.02868.x] [PMID: 16611267]

[59] Dobson AD, Cotter PD, Dobson AD, Cotter PD, Ross D, Hill C. Bacteriocin Production: a Probiotic Trait? Appl Environ Microbiol 2012; 78: 11-6.
[http://dx.doi.org/10.1128/AEM.05576-11]

[60] Figueroa-González I, Quijano G, Ramírez G, Cruz-Guerrero A. Probiotics and prebiotics--perspectives and challenges. J Sci Food Agric 2011; 91(8): 1341-8.
[http://dx.doi.org/10.1002/jsfa.4367] [PMID: 21445871]

[61] de Vrese M, Stegelmann A, Richter B, Fenselau S, Laue C, Schrezenmeir J. Probiotics--compensation for lactase insufficiency. Am J Clin Nutr 2001; 73(2) (Suppl.): 421S-9S.
[PMID: 11157352]

[62] Vonk RJ, Gerlof AR, Reckman, Harmsen JM, Priebe MG. Probiotics and Lactose Intolerance InTech 2012; 150-60.

[63] Coconnier MH, Lievin V, Hemery E, Servin AL. Antagonistic activity against *Helicobacter* infection *in vitro* and *in vivo* by the human *Lactobacillus acidophilus* strain LB. Appl Environ Microbiol 1998; 64(11): 4573-80.
[PMID: 9797324]

[64] Felley CP, Corthésy-Theulaz I, Rivero JL, *et al.* Favourable effect of an acidified milk (LC-1) on *Helicobacter pylori* gastritis in man. Eur J Gastroenterol Hepatol 2001; 13(1): 25-9.
[http://dx.doi.org/10.1097/00042737-200101000-00005] [PMID: 11204805]

[65] Pantoflickova D, Corthésy-Theulaz I, Dorta G, *et al.* Favourable effect of regular intake of fermented milk containing *Lactobacillus johnsonii* on *Helicobacter pylori* associated gastritis. Aliment Pharmacol Ther 2003; 18(8): 805-13.
[http://dx.doi.org/10.1046/j.1365-2036.2003.01675.x] [PMID: 14535874]

[66] Tong JL, Ran ZH, Shen J, Zhang CX, Xiao SD. Meta-analysis: the effect of supplementation with probiotics on eradication rates and adverse events during *Helicobacter pylori* eradication therapy. Aliment Pharmacol Ther 2007; 25(2): 155-68.
[http://dx.doi.org/10.1111/j.1365-2036.2006.03179.x] [PMID: 17229240]

[67] Sheu BS, Cheng HC, Kao AW, *et al.* Pretreatment with *Lactobacillus*- and *Bifidobacterium*-containing yogurt can improve the efficacy of quadruple therapy in eradicating residual *Helicobacter pylori* infection after failed triple therapy. Am J Clin Nutr 2006; 83(4): 864-9.
[PMID: 16600940]

[68] Hamilton-Miller JM. The role of probiotics in the treatment and prevention of *Helicobacter pylori* infection. Int J Antimicrob Agents 2003; 22(4): 360-6.
[http://dx.doi.org/10.1016/S0924-8579(03)00153-5] [PMID: 14522098]

[69] Dajani AI, Abu Hammour AM, Yang DH, *et al.* Do probiotics improve eradication response to *Helicobacter pylori* on standard triple or sequential therapy? Saudi J Gastroenterol 2013; 19(3): 113-20.
[http://dx.doi.org/10.4103/1319-3767.111953] [PMID: 23680708]

[70] Hirayama K, Rafter J. The role of probiotic bacteria in cancer prevention. Microbes Infect 2000; 2(6): 681-6.
[http://dx.doi.org/10.1016/S1286-4579(00)00357-9] [PMID: 10884619]

[71] Rafter J. Lactic acid bacteria and cancer: mechanistic perspective. Br J Nutr 2002; 88 (Suppl. 1): S89-94.
[http://dx.doi.org/10.1079/BJN2002633] [PMID: 12215183]

[72] Tavan E, Cayuela C, Antoine JM, Cassand P. Antimutagenic activities of various lactic acid bacteria against food mutagens: heterocyclic amines. J Dairy Res 2002; 69(2): 335-41.
[http://dx.doi.org/10.1017/S002202990200540X] [PMID: 12222811]

[73] Lo PR, Yu RC, Chou CC, Huang EC. Determinations of the antimutagenic activities of several probiotic bifidobacteria under acidic and bile conditions against benzo[a]pyrene by a modified Ames test. Int J Food Microbiol 2004; 93(2): 249-57.
[http://dx.doi.org/10.1016/j.ijfoodmicro.2003.11.008] [PMID: 15135962]

[74] Chiu YH, Hsieh YJ, Liao KW, Peng KC. Preferential promotion of apoptosis of monocytes by Lactobacillus casei rhamnosus soluble factors. Clin Nutr 2010; 29(1): 131-40.
[http://dx.doi.org/10.1016/j.clnu.2009.07.004] [PMID: 19665262]

[75] Iyer C, Kosters A, Sethi G, Kunnumakkara AB, Aggarwal BB, Versalovic J. Probiotic *Lactobacillus reuteri* promotes TNF-induced apoptosis in human myeloid leukemia-derived cells by modulation of NF-kappaB and MAPK signalling. Cell Microbiol 2008; 10(7): 1442-52.
[http://dx.doi.org/10.1111/j.1462-5822.2008.01137.x] [PMID: 18331465]

[76] Cenci G, Rossi J, Trotta F, Caldini G. Lactic acid bacteria isolated from dairy products inhibit genotoxic effect of 4-nitroquinoline-1-oxide in SOS-chromotest. Syst Appl Microbiol 2002; 25(4): 483-90.
[http://dx.doi.org/10.1078/07232020260517607] [PMID: 12583707]

[77] Geier MS, Butler RN, Howarth GS. Probiotics, prebiotics and synbiotics: a role in chemoprevention for colorectal cancer? Cancer Biol Ther 2006; 5(10): 1265-9.
[http://dx.doi.org/10.4161/cbt.5.10.3296] [PMID: 16969130]

[78] Kim Y, Oh S, Yun HS, Oh S, Kim SH. Cell-bound exopolysaccharide from probiotic bacteria induces autophagic cell death of tumour cells. Lett Appl Microbiol 2010; 51(2): 123-30.
[PMID: 20536712]

[79] Lee JW, Shin JG, Kim EH, *et al.* Immunomodulatory and antitumor effects *in vivo* by the cytoplasmic fraction of *Lactobacillus casei* and *Bifidobacterium longum.* J Vet Sci 2004; 5(1): 41-8.
[PMID: 15028884]

[80] Baldwin C, Millette M, Oth D, Ruiz MT, Luquet FM, Lacroix M. Probiotic *Lactobacillus acidophilus* and *L. casei* mix sensitize colorectal tumoral cells to 5-fluorouracil-induced apoptosis. Nutr Cancer 2010; 62(3): 371-8.
[http://dx.doi.org/10.1080/01635580903407197] [PMID: 20358475]

[81] Altonsy MO, Andrews SC, Tuohy KM. Differential induction of apoptosis in human colonic carcinoma cells (Caco-2) by *Atopobium*, and commensal, probiotic and enteropathogenic bacteria: mediation by the mitochondrial pathway. Int J Food Microbiol 2010; 137(2-3): 190-203.
[http://dx.doi.org/10.1016/j.ijfoodmicro.2009.11.015] [PMID: 20036023]

[82] Daniluk U. Probiotics, the New Approach for Cancer Prevention and/or Potentialization of Anti-Cancer Treatment? J Clin Exp Oncol 2012; 1: 2.
[http://dx.doi.org/10.4172/2324-9110.1000e105]

[83] Zuang G, Xiao-Ming L, Qiu-Xiang Z, *et al.* Research Advances With Regards to Clinical Outcome and Potential Mechanisms of the Cholesterol-Lowering Effects of Probiotics. Clin Lipidol 2012; 7: 501-7.
[http://dx.doi.org/10.2217/clp.12.40]

[84] Lambert JM, Bongers RS, de Vos WM, Kleerebezem M. Functional analysis of four bile salt hydrolase and penicillin acylase family members in *Lactobacillus plantarum* WCFS1. Appl Environ Microbiol 2008; 74(15): 4719-26.
[http://dx.doi.org/10.1128/AEM.00137-08] [PMID: 18539794]

[85] Pereira DI, Gibson GR. Cholesterol assimilation by lactic acid bacteria and bifidobacteria isolated from the human gut. Appl Environ Microbiol 2002; 68(9): 4689-93.
[http://dx.doi.org/10.1128/AEM.68.9.4689-4693.2002] [PMID: 12200334]

[86] Liong MT, Shah NP. Effects of a *Lactobacillus casei* synbiotic on serum lipoprotein, intestinal microflora, and organic acids in rats. J Dairy Sci 2006; 89(5): 1390-9.
[http://dx.doi.org/10.3168/jds.S0022-0302(06)72207-X] [PMID: 16606710]

[87] Liong MT, Shah NP. Acid and bile tolerance and cholesterol removal ability of *lactobacilli* strains. J Dairy Sci 2005; 88(1): 55-66.
[http://dx.doi.org/10.3168/jds.S0022-0302(05)72662-X] [PMID: 15591367]

[88] Lye HS, Rusul G, Liong MT. Mechanisms of Cholesterol Removal by Lactoballi Under Conditions That Mimic the Human Gastrointestinal Tract. Int Dairy J 2010; 20: 169-75.
[http://dx.doi.org/10.1016/j.idairyj.2009.10.003]

[89] Lye HS, Rusul G, Liong MT. Removal of cholesterol by lactobacilli via incorporation and conversion to coprostanol. J Dairy Sci 2010; 93(4): 1383-92.
[http://dx.doi.org/10.3168/jds.2009-2574] [PMID: 20338415]

[90] De Preter V, Vanhoutte T, Huys G, *et al.* Effects of *Lactobacillus casei* Shirota, *Bifidobacterium breve*, and oligofructose-enriched inulin on colonic nitrogen-protein metabolism in healthy humans. Am J Physiol Gastrointest Liver Physiol 2007; 292(1): G358-68.
[http://dx.doi.org/10.1152/ajpgi.00052.2006] [PMID: 16990449]

[91] Baquerizo Nole KL, Yim E, Keri JE. Probiotics and prebiotics in dermatology. J Am Acad Dermatol 2014; 71(4): 814-21.

[http://dx.doi.org/10.1016/j.jaad.2014.04.050] [PMID: 24906613]

[92] Yazdanbakhsh M, Kremsner PG, van Ree R. Allergy, parasites, and the hygiene hypothesis. Science 2002; 296(5567): 490-4. [http://dx.doi.org/10.1126/science.296.5567.490] [PMID: 11964470]

[93] Kalliomäki M, Kirjavainen P, Eerola E, Kero P, Salminen S, Isolauri E. Distinct patterns of neonatal gut microflora in infants in whom atopy was and was not developing. J Allergy Clin Immunol 2001; 107(1): 129-34. [http://dx.doi.org/10.1067/mai.2001.111237] [PMID: 11150002]

[94] Özdemir Ö. Gut flora development in infancy and its effect on immune system. çocuk Enf Derg (J Pediatr Inf) 2009; 3: 202-3.

[95] Gueimonde M, Kalliomäki M, Isolauri E, Salminen S. Probiotic intervention in neonates--will permanent colonization ensue? J Pediatr Gastroenterol Nutr 2006; 42(5): 604-6. [http://dx.doi.org/10.1097/01.mpg.0000221897.45910.d3] [PMID: 16707993]

[96] Vandenplas Y, Huys G, Daube G. Probiotics: an update. J Pediatr 2014. [PMID: 25458874]

[97] Isolauri E, Arvola T, Sütas Y, Moilanen E, Salminen S. Probiotics in the management of atopic eczema. Clin Exp Allergy 2000; 30(11): 1604-10. [http://dx.doi.org/10.1046/j.1365-2222.2000.00943.x] [PMID: 11069570]

[98] Vasiljevic T, Shah NP. Probiotics—From Metchnikoff to bioactives. Int Dairy J 2008; 18: 714-28. [http://dx.doi.org/10.1016/j.idairyj.2008.03.004]

[99] Tuohy KM, Probert HM, Smejkal CW, Gibson GR. Using probiotics and prebiotics to improve gut health. Drug Discov Today 2003; 8(15): 692-700. [http://dx.doi.org/10.1016/S1359-6446(03)02746-6] [PMID: 12927512]

[100] Garcia Vilela E, De Lourdes De Abreu Ferrari M, Oswaldo Da Gama Torres H, *et al.* Influence of Saccharomyces boulardii on the intestinal permeability of patients with Crohn's disease in remission. Scand J Gastroenterol 2008; 43(7): 842-8. [http://dx.doi.org/10.1080/00365520801943354] [PMID: 18584523]

[101] Gionchetti P, Rizzello F, Venturi A, *et al.* Oral bacteriotherapy as maintenance treatment in patients with chronic pouchitis: a double-blind, placebo-controlled trial. Gastroenterology 2000; 119(2): 305-9. [http://dx.doi.org/10.1053/gast.2000.9370] [PMID: 10930365]

[102] Pronio A, Montesani C, Butteroni C, *et al.* Probiotic administration in patients with ileal pouch-anal anastomosis for ulcerative colitis is associated with expansion of mucosal regulatory cells. Inflamm Bowel Dis 2008; 14(5): 662-8. [http://dx.doi.org/10.1002/ibd.20369] [PMID: 18240282]

[103] Nanda Kumar NS, Balamurugan R, Jayakanthan K, Pulimood A, Pugazhendhi S, Ramakrishna BS. Probiotic administration alters the gut flora and attenuates colitis in mice administered dextran sodium sulfate. J Gastroenterol Hepatol 2008; 23(12): 1834-9. [http://dx.doi.org/10.1111/j.1440-1746.2008.05723.x] [PMID: 19120873]

[104] Haller D, Antoine JM, Bengmark S, Enck P, Rijkers GT, Lenoir-Wijnkoop I. Guidance for substantiating the evidence for beneficial effects of probiotics: probiotics in chronic inflammatory

bowel disease and the functional disorder irritable bowel syndrome. J Nutr 2010; 140(3): 690S-7S. [http://dx.doi.org/10.3945/jn.109.113746] [PMID: 20107148]

CHAPTER 5

Immunotherapy

Solange Gil[a,*,+], Dias J.[a,+], Lourenço A.M.[a,+], Leal R.[b,+], Tavares L.[a,+]

[a] *Centro de Investigação Interdisciplinar em Sanidade Animal (CIISA), Faculdade de Medicina Veterinária, Technical University of Lisbon (TULisbon), Av. Universidade Tècnica, 1300-477 Lisbon, Portugal*

[b] *Centre Hospitalier Vétérinaire Fregis, 43 Av. Aristide Briand 94110, Arcueil, France*

Abstract: The relationship between a host and a pathogen is dynamic, and its outcome depends on the virulence of the invader and the relative degree of resistance or susceptibility of the host at that particular occasion. Fortunately, only an infinitesimal minority of microorganisms are able to avoid the host defenses and eventually cause disease. The host has evolved multifaceted strategies for defending itself against invasion, and pathogens have evolved their own strategies of counter-attack for host defenses pointing us to the co-evolution of "defense" and "attack" mechanisms. Many factors determine the outcome of the bacterium-host relationship. In this battle, the development of antibiotics was a game-changing turn. Unfortunately, more and more strains of pathogenic bacteria have become antibiotic resistant. One of this century's greatest medical challenges is the rapid emergence of multidrug-resistant pathogens. Discovery of new antimicrobial classes and alternative therapeutic options would be especially welcomed in this era. The fact that some bacterial infections might no longer be successfully treated with antibiotics, combined with an increasing population density and mobility, justifies the urgent demands for the development of novel treatments, of which immunotherapies are considered most promising. Immuno modulatory regimens offer an attractive approach as they often have fewer side effects than existing drugs, including less potential for creating resistance in microbial diseases. Nowadays, treating emerging chronic and multidrug-resistant infectious, autoimmune and oncological diseases through reinforcement of the immune system became a major priority.

Keywords: Antibiotics, Antibodies, Autoimmune diseases, Bacteria, Bacterial infectivity, Cancer, Evasion immune system, Growth Factors, Host resistance, Host defences, Immune response, Immune system, Immunomodulators, Immuno

[*] **Corresponding author Solange Gil:** ICentro de Investigação Interdisciplinar em Sanidade Animal (CIISA), Faculdade de Medicina Veterinária da Universidade de Lisboa, Lisboa, Portugal; E-mail:solange@fmv.ulisboa.pt

[+] **Equally contributed**

Manuela Oliveira and Isa Serrano (Eds)

therapy, Interferons, Interleukins, Multiresistant Infection, Passive Immuno therapy, Vaccines, Virulence.

1. BACTERIAL INFECTIVITY, IMMUNE RESPONSE AND MULTIRESISTANT INFECTIONS : NEW PERSPECTIVES

Bacterial infectivity results from a disturbance in the balance between bacterial virulence and host resistance. The "objective" of bacteria is to multiply rather than to cause disease; it is in the best interest of the bacteria not to kill the host. However, sometimes the story is not that simple.

1.1. Introduction

'good fences make good neighbours'

This section is about a fine tuned dance practiced over and over again through evolutionary times between a pathogen and its host, in a very Darwinistic way. This is the story of pathogens that are usually smaller than a human cell looking for resource-rich environment in which to survive and reproduce, and of their host fighting them with defense mechanisms and experience accumulated over hundred millions of years of evolution [1].

Infection is the invasion of the host by microorganisms, which then reproduce in close association with the host's tissues. Bacteria can cause a multitude of different infections, ranging in severity from unapparent to fulminating [2].

The ability of bacteria to cause disease reflects its relative pathogenicity. On this basis, bacteria can be organized into three major groups. Primary pathogens are considered to be probable agents of disease (*e.g.*, when the cause of diarrheal disease is identified by the laboratory isolation of *Campylobacter* spp. from feces). Opportunistic pathogens are those who colonize the body with no ill effect for most of the time but which can cause illness if the body's defenses are undermined. *Staphylococcus aureus*, a bacterium that colonizes human skin is also a common cause of pimples and can even cause food poisoning. *Escherichia coli* does no harm if it stays in the intestinal tract, but it can cause disease if it gains access to the urinary tract or the bloodstream. Finally, some bacteria, such as *Lactobacillus acidophilus* for man, are considered to be nonpathogens, because they rarely or never cause disease in their host. Some nonpathogens are beneficial for the host and some have no effect. Nevertheless, their categorization as nonpathogens is in no way final, in part due to plasticity of bacteria. In fact, some bacteria previously considered to be nonpathogens are now known to cause disease.

Virulence is the measure of the pathogenicity of an organism. The degree of virulence is related directly to the ability of the organism to cause disease despite host resistance mechanisms [2]; it relies upon many variables such as the number of infecting bacteria, the route of entry into the body, innate and adaptive host defense mechanisms, and inherent virulence factors of the bacteria. Therefore, the determinants of virulence of a pathogen are any of its genetic or biochemical/structural features that facilitate its process to produce disease in an individual [2].

The relationship between a host and a pathogen is dynamic, and its outcome depends on the virulence of the bacteria and the relative degree of resistance or susceptibility of the host at that particular occasion.

Virulence can be measured experimentally by determining the number of bacteria required to cause animal death, illness, or lesions after the bacteria are administered by a designated route. Calculations of a lethal dose affecting 50 percent of a population of animals (LD_{50}) or an effective dose causing a disease symptom in 50 percent of a population of animals (ED_{50}) are useful in comparing the relative virulence of different bacteria [3].

Pathogenesis refers both to the mechanism of infection and to the mechanism by which disease develops. Here, we focus on the first aspect.

Fortunately for us, only an infinitesimal minority of bacteria are able to avoid the host defenses and eventually cause disease. In order to comprehend the mechanisms by which pathogenic bacteria cause disease, let us briefly mention some of the obstacles those small numbers of bacteria must overcome.

1.2. Host Resistance

The skin is the body's first defense against the microbial world and has several hidden weapons to be successful. Firstly, it forms a tight, impenetrable barrier of epithelium and is constituted by cells with a rapid rate of replacement. Secondly, several antimicrobial products are present in the skin of many species (insects and other non-vertebrate organisms, amphibians, and more recently studied, human and canine skin) [4]. In fact, hundreds of naturally occurring peptide antibiotics (AMPs) have been lately discovered. These AMPs participate in the innate immune system response not only providing rapid first-line defense against infection but also acting as signaling molecules that activate host immune defense and tissue repair by promoting cytokine release, inflammation, angiogenesis, and reepithelialization. In human skin cathelicidins and defensins are the major AMPs' families. Alterations in the expression of AMPs have been associated with several diseases [5 - 8]. These ancestral molecules are nowadays, particularly with antibiotic resistance rising, attractive candidates for novel therapeutic approaches [9]. Third, a tight meshed network of antigen-presenting cells is established through the entire skin (skin-associated lymphoid tissue - SALT). This tailored immunological network is essential for maintenance of tolerance but also for the development of protective and controlled immune responses [10]. Finally, the skin is covered by commensal bacteria which can both directly protect humans and other species from pathogenic invaders and help the immune system maintain that delicate balance between effective protection and damaging inflammation [11 - 13]. Colonization is driven by the ecology of the skin, which is extremely changeable depending on topographical location, endogenous host and exogenous environmental factors. The cutaneous innate and adaptive immune responses can modulate the skin microbiota, but the microbiota also aids in educating the immune system [14]. For instance, *Staphylococcus epidermidis*, a skin commensal bacterium, has been shown to induce cutaneous interferon (IFN)-γ- and interleukin (IL)-17A-producing T cells [10]. In addition dysbiosis, which can occur as a result of, for example, a chronic disease or the incorrect use of antibiotic supplementation, has been shown to even alter wound repair, reinforcing the importance of commensal microbiota for normal function of the skin [15]. Therefore, nowadays the skin immune system should be considered a cooperative of elements from the host to microbes acting in a mutualistic relationship [16]. This concept is an exciting, novel, game changing one and much will still be written on this subject.

Nevertheless, the skin is easily damaged and upon breach it can expose the soft tissues underneath rendering them vulnerable for infection. As a matter of fact, before World War II (1939-1945) the most common cause of death during the wartime was bacterial infections and not enemy bullets [17].

In addition to the skin barrier, mucous membranes of the respiratory, gastrointestinal, and urogenital systems represent other portals through which bacteria can gain access to the body. They are topologically ‘‘inside’’ the body, but exposed constantly to the outer environment. Unlike the many dead layers found in skin, mucosal epithelia are comprised of only one epithelial layer, continuously bathed in fluids and with warm environment, offering therefore ideal conditions for the growth of bacteria [18]. To protect against bacterial colonization, these vulnerable epithelia have evolved a formidable array of chemical and physical barriers. Let's just mention that there is some noticeable mimicry regarding the way the most vulnerable parts of the body defend against pathogens.

The mucosal epithelial cells also divide rapidly, and as the cells mature, they shed into the lumen, thus bacteria that manage to reach and colonize a mucosal surface are constantly being eliminated. Most mucosal surfaces are protected by a normal resident microbiota. Exceptions are the uterus and upper female genital tract and the urinary tract [18]. As with the skin, there is increasing evidence that not only the gut microbiome ensures competition for pathogens to which they are exposed but also provides the microbial interactions necessary for the development of regulated pro- and anti-inflammatory signals that promote both adequate defense and immunological tolerance [19 - 21]. Regulation of microflora composition (*e.g.* by probiotics, prebiotics or symbiotics) influences the development of mucosal and systemic immunity and may play a role in prevention and treatment of some diseases [22].

The human gut contains, in the assortment of 1000 bacterial species, 100-fold more genes than the human genome and a huge potential of antigenic diversity for immune stimulation. The function of the gut microbiome in human health and disease is becoming more elucidated thanks to high throughput sequencing technologies in addition to parallel recent developments in nongenomic techniques [23 - 25].

An essential protection of many internal epithelia is mucus as it traps bacteria, prevents them from reaching the surfaces of the epithelial cell, and facilitates their elimination. Additionally, another protective role of mucus is to bind proteins that have antibacterial activity such as lysozyme, lactoferrin, and lactoperoxidase [18]. These substances either kill bacteria or restrict their growth.

AMPs, the ancient or primitive defense mechanisms are present in virtually every multi-cellular organism and in mucous membranes as well. These antimicrobial peptides can deflect ingested pathogens but also shape and control the orchestra of commensal population. The most prominent families of proteins featuring AMPs in the intestinal tract of humans are represented by defensins, cathelicidins (*e.g.* LL-37), C-type lectins ribonucleases (RNases), and S100 proteins (example: calprotectin).

As with the skin, mucosal surfaces have an underlying population of phagocytic cells and immune cells. This mucosal defense system is called the mucosa-associated-lymphoid tissue (MALT). Their main function is to make secretory immunoglobulin A (SIgA), an antibody class that is secreted into mucus on the apical side of the epithelium [18]. SIgA's multifaceted roles in controlling inflammation, regulating immune responses, and mounting defenses to enteric pathogens are only beginning to be understood. There is emergent evidence that indicates that SIgA influences the composition of the gut microbiota, promotes the uptake and delivery of antigens from the intestinal lumen to dendritic cells subsets located in gut-associated lymphoid tissue (GALT), and influences inflammatory responses associated with the uptake of highly pathogenic bacteria [26].

During the normal course of life, individuals develop local antibodies specific for a variety of bacteria that colonize mucosal surfaces. Other special defenses of the intestinal tract include the extremely acidic environment of the stomach; the bile salts with their detergent-like properties that help to disrupt bacterial membranes, especially those of gram negative bacteria, and the rapid flow of contents through the small intestine [18].

A broad mechanism of restricting the growth of bacteria that penetrate the skin and mucous membranes is the competition for iron. Also, the phagocytic cells of the body tour the blood and tissues for strange substances, including bacteria. This mission is assumed predominantly by polymorphonuclear neutrophils; however, monocytes, macrophages, and eosinophils also contribute. After phagocytosis, bacterial cells usually are killed except if their number is excessive or they possess virulence factors that allow them to survive the lysosomal enzymes and acidic pH. Complement, an enzymatic system of serum proteins, is considered as part of the innate immunity because of its role in inflammation, phagocytosis and bacterial killing. Complement may be activated by bacterial invasion, but also by reactions between antigens and antibodies, and therefore, it may play a role in adaptive immunity as well. During these interactions T cells, and B cells, specific antibody responses, and/or cell-mediated immunity develop to protect against reinfection [2].

1.3. Bacterial Strategies for Evading, Surviving Host Defense Systems and Create Infection

It is a common saying that "it takes two to tango". Partners in dance adapt, evolve together taking into account each other moves and signals. Time is something our partners in evolution had in order to achieve a remarkable understanding of each other resulting in a much paired existence. Just as the host has evolved multifaceted strategies for defending itself against bacterial invasion, bacteria have evolved their own strategies of counter-attack for host defenses pointing us to the co-evolution of "defense" and "attack" mechanisms. The ability of pathogenic bacteria to cause infection and illness depends on the structural, biochemical, and genetic characteristics that contribute to a microbe's pathogenicity [18].

Two broad qualities underline the ability of bacteria to cause infection: invasiveness and virulence. The initial step will depend on the mechanisms for colonization (adherence and initial multiplication) and evasion of the host immune response. Bacteria have found ways to:

- **Penetrate intact skin.** Although they cannot do it unaided, for example *Erlichia canis*-agent of Ehrlichiosis, a disease most commonly affecting canine species, has created its own skin trauma opportunities by colonizing biting arthropods

(*Rhipicephalus sanguineus*); also, staphylococcal exfoliative toxins, act as "molecular scissors" to facilitate percutaneous bacterial invasion of mammalian skin by cleavage of keratinocyte cell–cell adhesion molecules [27];

- **Subvert the mucin barrier.** Some pathogens can penetrate the mucus barrier physically, through flagella-mediated motility or by enzymatic degradation of the mucus. As result, both pathogens and commensal bacteria can reach the epithelium [28];
- **To produce IgA proteases.** As mentioned before secretory immunoglobulin A (SIgA) is the principal immunologic defense of respiratory and other mucosal surfaces in the body. Cleavage of SIgA by bacterial proteases has been shown and might render it immunologically inactive and thus contribute to the development of infections [29];
- **Resist antibacterial peptides-** Several studies are showing different strategies of natural or induced resistance to AMPs among bacterial species. However all resistance mechanisms have its flaws and bacterial have never succeeded in becoming entirely insensitive to broad range AMPs [30]. *Staphylococcus aureus* and *Salmonella enterica* are among bacterial pathogens that have developed countermeasures in order to reduce the effectiveness of cationic AMP's. These include the repulsion of the AMP's by dropping the net negative charge of the bacterial cell envelope; expelling of AMPs through energy-dependent pumps; altering membrane fluidity; and cleaving AMPs with proteases. Mutants susceptible to AMPs were found to be more efficiently inactivated by phagocytes and are virulent-attenuated, signifying that AMP resistance influences bacterial infection [30 - 32].
- **Acquire iron-** In order to circumvent nutritional immunity (sequestering essential nutrients) bacterial pathogens have evolved elegant systems that allow the acquisition of iron. This is accomplished among others by production of siderophorus, compounds that chelate iron with high affinity; proteases that degrade transferrin or lactoferrin in order to release iron; acquisition of iron directly from hemoglobolin or heme; production of toxic factors, such as hemolisins, that kill host cells and consequently release iron; or the most extreme adaptation, the iron abstinence strategy (*Borrelia burgdorferi* the bacterium that causes Lyme diseases can live without iron) [33, 34]. Some of the strategies depended on the iron limitation. One exceptionally resourceful

bacterium is *S. aureus* with several molecular mechanisms of iron acquisition being described. For example, under severe iron-limiting conditions *S. aureus* increases fermentative metabolism, resulting in increased production of lactate which lowers the pH of the microenvironment and the affinity of transferrin for iron [35].

- **Adhere-** Virtually all known bacterial pathogens have ways of attaching themselves firmly to host cells. The main strategies bacteria use to attach themselves to host cells involve adhesion structures like pili, fimbriae, non-fimbriae adhesions and others [18].
- **Evade the host immune response-** Some of the strategies used for this purpose are avoiding complement and phagocytosis (*e.g.* by producing a capsule, a network of polymers that covers the surface of the bacterium); mimicking the host, for example some bacterial capsules have a composition that resemble host polysaccharides, such as those consisting of hyaluronic acid an extracellular matrix polysaccharide produced by *S. pyogenes*, this way they are not immunogenic meaning no antibodies will be produced against them [36]; producing toxins that kill phagocytes, inhibit their migration or prevent activation; hiding from phagocytes by producing *invasins* (bacterial surface proteins that provoke phagocytic ingestion of the bacterium by host cells allowing bacteria like *Yersinia pseudotuberculosis* to liveinside the host cell); surviving phagocytosis (for example by escaping from the phagosome); evading the host antibody response (the bacterium to be mistaken for part of the host itself); or enhancing tissue penetration and dissemination by producing the so called "spreading factors", such as hyaluronidases [18].

The key to the development of disease in many bacterial infections is expression of a bacterial toxin. Bacterial toxins damage the host at the site of bacterial infection or distant from the site of infection. Bacterial toxins were the first virulence factors to be identified and studied in detail. Some have functions we already mentioned, for example they may act at the site of colonization and play a role in invasion, or by killing neutrophils and macrophages they are able to protect bacteria from phagocytic cells. But their repertoire is very complex, growing, and sometimes their function more subtle, such as modulation of the signaling pathways that regulate cytokine production. The widespread production of toxins and the number of toxins that are associated with human and animal disease, lead to an inevitable conclusion that toxins are important for the bacterial survival and multiplication [18]. In fact, toxins detrimental capacity to humans and their versatility makes them potential tools of bioterrorism. Toxins have properties of both of biological and chemical weapons [37, 38].

Invasive bacteria injure the host through the production of extracellular enzymes that damage host tissue, or through the modulation of the host response system. In contrast, bacterial pathogens that damage the host through the action of toxins are often non-invasive and have limited capacity to disseminate in the host [39].

Bacteria may produce two types of toxins called exotoxins and endotoxins. Exotoxins are released from bacterial cells and may act at remote tissue sites distant from the site of bacterial growth. Major categories include cytotoxins, neurotoxins, and enterotoxins. Probably the most famous toxin is produced by *Clostridium botulinum*, botulinum toxin. Botulinum toxin inhibits muscle contraction by inhibiting the release of a neuro-transmitter hormone, acetylcholine. Nevertheless, it has been converted into a cosmetic product, Botox, which is used to avoid contraction of muscles causing the formation of wrinkles. Endotoxins are cell-associated substances. However, endotoxins may be released from growing bacterial cells and cells that are lysed as a result of effective host defense or of the activities of certain antibiotics (*e.g.* penicillins) [18]. They can cause fever, changes in blood pressure, inflammation, lethal shock, and many other toxic events [2]. Hence, bacterial toxins, both soluble and cell-associated, may be transported by blood and lymph and cause cytotoxic effects at tissue sites remote from the original point of invasion or growth.

It is clear now that many factors determine the outcome of the bacterium-host relationship. In this battle the development of antibiotics was a game-changing turn. The discovery of penicillin in 1928 was followed by the discovery and commercial production of many other antibiotics. It was almost taken for granted that any bacterial infectious disease were curable by antibiotic therapy. But soon after the development of antibiotics, more and more strains of pathogens have become antibiotic resistant, and, with time, some have become resistant to many antibiotics and chemotherapeutic agents, the phenomenon of multidrug resistance.

Multidrug resistance in bacteria occurs by the accumulation of genes, each coding for resistance to a specific agent, and/or by the action of multidrug efflux pumps, each of which can pump out more than one drug type [40]. Infection by these resistant strains causes huge problems for human and animal population. As the World Health Organization (WHO) states "antimicrobial resistance threatens the effective prevention and treatment of an ever-increasing range of infections caused by bacteria, parasites, viruses and fungi", for example resistance to earlier generation antimalarial drugs is widespread in most malaria-endemic countries [41]. Without an urgent action the world can be heading towards a post-antibiotic era, in which common bacterial infections, which have been treatable for decades, can once again kill. Some of the facts pointed out by the WHO [41] are of major concern:

- Treatment failure to the drug of last resort for gonorrhoea – third-generation cephalosporins – has been confirmed in several countries;
- Resistance to fluoroquinolones, one of the most widely used antibacterial drugs for the oral treatment of urinary tract infections caused by *E. coli*, is widespread;
- Resistance to first-line drugs to treat infections caused by *Staphlylococcus aureus* is also widespread, with MRSA (methicillin-resistant *Staphylococcus aureus*) infections becoming more common every day. In accordance, the canine affecting species, *Staphylococcus pseudintermedius,* is also developing serious resistance [42];
- Resistance to carbapenem antibiotics, the last resort treatment for life-threatening infections caused by common intestinal bacteria, has spread to all regions of the world.

It is however important to understand that multidrug resistance is not synonymous of virulence. For example humans can be colonized with MRSA meaning they carry the bacteria on their skin or in their nose but have no symptoms [43 - 45]. This bacterium will only cause disease if there is some condition, like a cut, that it will give the chance the bacteria needs to take advantage and cause disease. Then, antibiotic options will be limited.

The fact that antibiotics cannot kill particular bacteria does not mean the host, with all its resources, given adequate conditions (for instance management of the primary disease that predisposed for the bacterial infection in the first hand *e.g.* atopic dermatitis), may not. Obviously, any infection, but in particular those with few therapeutic options, pose a much more serious problem for immuno compromised individuals. As the growing prevalence of antimicrobial resistance in major pathogens is outpacing discovery of new antimicrobial classes, alternative therapeutic options that could help individuals deal with those infections would be especially welcomed in this era.

Vaccines (pathogen-specific immunotherapy) might be a good strategy for countering the rise of drug resistance. At the moment, vaccines are defined as nontoxic or greatly attenuated antigens that are injected, ingested, or inhaled to induce the specific protective host immune defenses. Vaccine-mediated protection bacterial disease in general, and multiresistant infections in particular, holds promise to have significant health benefits. The use of effective vaccines has the potential to enhance antibiotic efficacy and mitigate resistance, by reducing and allowing a more selected, pinpoint use of these drugs [46]. Unfortunately, nowadays vaccines for many drug-resistant pathogens remain undiscovered or have limited efficacy, and passive immunization strategies are greatly unsuccessfully as well [47]. For example, for *Staphylococcus aureus* recurring infections there is a limited success of vaccines containing *S. aureus* antigens implying that native antigens induce immune responses insufficient for optimal efficacy. There is a need to apply novel vaccine strategies to high-priority pathogens. Some authors propose a strategy to target vaccines against drug-resistant pathogens, by using resistance-conferring proteins as antigens in multicomponent vaccines [48]. While other authors support the novel concept of an anti-virulence, toxin-based vaccine intended for prevention of clinical invasive disease, rather than achieving sterile immunity [49]. As an alternative, there are those who advocateapplying convergent immunity, where antigens from non-target organisms that contain homologs epitopes found in the target organism are applied in vaccines. Convergent immunity has a potential to result in a protective efficacy not usually elicited by native antigens from a target pathogen [46]. This could be used to create innovative vaccines targeting among others *Staphylococcus aureus*.

Optimally efficacious vaccines and immunotherapeutics protective against bacterial pathogens likely rely upon a more broad understanding of the critical immunoprotective mechanisms targeting the organisms; innovative design (including the use of convergent immunity) which optimize efficacy and promote immunity that would not otherwise be induced by natural antigens; immunization protocols and strategies that induce powerful and harmonized immune responses; strategic adjuvants; and combinations of all the above in order to offer protection against the ultimate target pathogen.

In general terms, when infection with pathogenic bacteria results in disease it means that host resistance mechanisms, such has natural barriers and specific immunity against that pathogen, failed. The reasons underlying this failure might be multiple and diverse, occurring both at the host or pathogen levels, ranging from inappropriate antigenic presentation, lack of competent stimuli, too a high pathogen load or use of efficacious evasion strategies. In several situations, manipulation of the host immune system, through strategies to reinforce or enhance specific immune responses (including antigenic presentation using a different route, induction or inoculation of immune mediators such as cytokines), can be an efficacious alternative to the use of antibiotics. These approaches can be generally described as immunotherapies and some examples of success of these strategies will be described below.

2. IMMUNOTHERAPY

2.1. Introduction

Immunotherapy can be described as the treatment that promotes the stimulation or reestablishment of the immune system ability to fight infection and disease.

Empirical use of immunotherapy strategies has been described since the ancient times in Egypt and China where "sniffing" of crusts from small pox lesions was used to prevent disease 3-2000 years BC. Other examples of immunotherapy in history include the description of a Dorset farmer Benjamin Jesty (1774) who inoculated his wife and two sons with pus from the teat of a cow suffering from cowpox, using his wife's knitting needle as an empirical vaccination [50]. However, it was only in 1796, when Edward Jenner performed the first scientific attempt to control an infectious disease by the deliberate use of vaccination that started one of the greatest revolutions in medical intervention [51]. Jenner applied experimental inoculations of cowpox to human volunteers and vaccination against smallpox quickly controlled the disease in Europe and North America [52].

Since then, human beings have benefited from immunotherapy for more than two centuries and many of the devastating infectious diseases such as measles, diphtheria, smallpox, and pertussis have been contained, especially in industrialized nations, because of the development and widespread distribution of safe, effective, and affordable vaccines [53]. Nowadays, treating emerging chronic and multidrug-resistant infectious, autoimmune and oncological diseases through harnessing the immune system became also a major priority [51, 54].

One of this century’s greatest medical challenges is the rapid emergence of multidrug-resistant pathogens, especially Gram-negative bacteria and myco-bacteria. The fact that bacterial infections might no longer be successfully treated with antibiotics, combined with an increasing population density and mobility, urgently demands for the development of novel treatments, where immuno-therapies are considered most promising [55]. Moreover, recent successes related to effective immunotherapy for human autoimmunity are benefiting almost a million patients with chronic diseases, such as rheumatoid arthritis and Crohn's disease, which are unresponsive to other treatments [56].

The concept of immunotherapy has been most appealing on the oncology research field because of the urge to find novel therapies capable of targeting non dividing cells. Generally, conventional chemotherapy yields high remission rates by eliminating actively dividing cells; however, a quiescent but lethal subpopulation of cells may be responsible for disease relapse. As patients are treated with increasingly toxic regimens, new approaches should be made safer by targeting tumor cells in a more specific way [57 - 59].

Novel immunotherapy strategies have only been available in the past 20 to 25 years, due to major advances in medical science and technology, but are now being increasingly used to treat a wide spectrum of diseases [56]. Immuno modulatory regimens offer an attractive approach as they often have fewer side effects than existing drugs, including less potential for creating resistance in microbial diseases.

There is a growing interest in the treatment of diseases such as cancer, autoimmune diseases and diseases caused by viruses using immunotherapy. This interest has been extended to bacterial diseases with particular focus on multi-resistant bacterial infections. The following paragraphs intend to review the main uses of immunotherapy in its different application areas.

2.2. Applications of Immunotherapy

Immunotherapies are designed to treat disease by inducing, enhancing or suppressing an immune response. In the case of some diseases such as cancer, the immune system seems to be ineffective in eliminating the "abnormal" transformed cells, and this therapeutic option attempts to strengthen the immune system in its fight against cancer.

Furthermore, in some viruses, the organism does not produce enough natural anti-viral molecules. For example, both hepatitis B and hepatitis C are treated with interferon, often in combination with other antiviral drugs, and some of the patients have a sustained virological response and can eliminate hepatitis virus [60].

Ultimately, in the case of an autoimmune disorder that occurs when the immune system attacks and destroys healthy self-tissues, immunotherapy represents an attempt to down regulate the overactive immune system [61, 62].

Immunotherapy can be regarded as specific or nonspecific. Specific immuno therapy aims to supply immune components against well-defined antigens (*i.e.* treatment with monoclonal antibodies) while non-specific therapy boosts all immune responses, for example, through the use of cytokines or Lymphokine Activated Killer (LAK) cells [51].

Additionally, immunotherapy can be classified in three main categories: passive, active and a new class called combination immunotherapy.

The adaptive immune system includes two groups of genes — B-cell and T-cell receptors (BCRs and TCRs, respectively) — that endure genetic recombination and express protein products that recognize antigens. Moreover B and T cells can adjust to, specifically bind, and eradicate pathogens or host cells presenting their antigens. Nowadays, therapeutic agents with engineered BCRs and TCRs (either monoclonal antibodies or transgenic T cells) targeting a chosen antigen can be developed and administered to the patient. Such immunotherapy is considered passive as it does not stimulate a patient's immune system, comprising the immune system components produced outside the organism given to patients to promote immunity against a disease [57].

An alternative approach is active immunotherapy that attempts to stimulate thebody's own immune system by presenting antigens that trigger an immune response. Subsequently, the patient is inoculated with a pool of potential target antigens allowing the immune system to select the most effective BCRs and TCRs [57].

Combination immunotherapy comprises both active and passive activity. While active and passive immunotherapies respectively mimic some parts of the immune system, combination immunotherapy has the potential to most closely resemble the natural immune system [63].

The active agents of immunotherapy are collectively called immunomodulators and include a diverse array of recombinant, synthetic and natural preparations.

2.3. Passive Immunotherapy

In 1888 Emile Roux and Alexandre Yersin isolated a soluble toxin from cultures of diphtheria. This provided the scientific basis for the work of Emil von Behring and Shibasabura Kitasato in Berlin at the Robert Koch Institute of Hygiene. In 1890, Emil von Behring and Shibasabura Kitasato published a landmark article showing that serum from actively immunized animals could neutralize toxic concentrations of toxin in other animals. They could also successfully cure children of diphtheria with horse antisera. Serotherapy was established as a treatment against diphtheria as well as tetanus toxin. Emil von Behring and Shibasaburo Kitasato were awarded the first Nobel Prize in Medicine in 1901 for their role in the discovery of passive immunotherapy [51, 52, 64].

Passive immunization, currently approved to treat infections caused by a number of toxin-producing bacteria, could be improved and extended to treat infections caused by highly drug-resistant bacteria, such as *S. aureus.*

2.4. Passive Immunotherapy: Antibodies

After preliminary successes in the late 1800s, sera from humans or animals were extensively used for viral and bacterial diseases prophylaxis and therapy. Ultimately, when antibiotics became largely available in the 1940s, serum therapy of the majority of bacterial infections was abandoned [64].

Nevertheless, polyclonal antibody formulations are still being used for some toxin-mediated infectious diseases and venomous bites. Serum immunoglobulin is also applied in the treatment of some viral diseases with few therapeutic options available and for pre or post-exposure prophylaxis [64].

Even though treatments with polyclonal antibody serum have shown clinical success in several cases, their application has been limited because of complications associated with toxicity, including a high risk of allergic reactions, lot- variation and undefined dosing. Moreover, a polyclonal serum usually contains a small quantity of active antigen-specific antibodies compared with the antibody total (1%). In fact, until 1970s it was impossible to produce large amounts of antibodies with a defined specificity [64].

Köhler and Milstein's discovery of hybridoma technology in 1975, allowed the unlimited production of monoclonal antibodies (mAbs) with chosen specificity, changing the paradigm for antibody therapy [64].

This cell fusion technique allowed the production of hybrids between myeloma cells and B lymphocytes. These hybrid lines could permanently grow in tissue culture and induce antibody production in mice using Jerne's hemolytic plaque assay [65].

The first therapeutic mAb was generated as OKT3 by Ortho Biotech in 1984 and consisted on a murine monoclonal antibody of the immunoglobulin against CD3 receptor found on human mature T cells and medullary thymocytes [66]. Unmodified mouse monoclonal antibodies exhibit a reduced therapeutic efficacy because they are immunogenic to other animal species and humans, have diminished *in vivo* survival and generally do not eliminate target cells efficiently due to their inability to fix human complement or to stimulate antibody-dependent cellular cytotoxicity (ADCC) [51].

To overcome this obstacles, a technique was developed that allows to clone the individual genes responsible for synthesizing human and murine immunoglobulins leading to the generation of antibody molecules having the variable region of mouse mAbs and the constant regions immunogenicity of human immunoglobulins. This method allows the selection of antibody specificity from among a large repertoire of murine mAbs, as well as selection of the human constant regions taking into account their biologic behavior [67].

An example of application in immunotherapy of this technique of producing chimeric antibodies is the launch of a product called Rituximab approved by the US Food and Drug Administration (FDA) to treat non-Hodgkin's lymphoma (NHL). It was the first approved mAb directed towards cancer cells. Rituximab is a chimeric antibody directed against the surface antigen CD20 on B lymphocytes, expressed on most B-cell NHL and subtypes of Acute Lymphatic Leukemias (ALL) and provided significant enhancements in the efficacy of treatment *versus* existing non-mAb therapies [51, 64, 68].

With the development of antibody engineering techniques and the better understanding of the antibody structure and function, it was possible to replace the hypervariable loops of a fully human antibody with the hypervariable loops of the murine antibody, decreasing even more the murine part of mAbs, using a method named complementarity-determining region grafting. These antibodies, called 'humanized', are 85–90% human and are less immunogenic than chimeric antibodies [68].

Finally, the increasing progress on the field of molecular biology allowed the development of transgenic animals and advanced techniques, including phage-display selection, leading to the engineering of fully human antibodies [64].

Antibody engineering allowed to clone entire repertoires of antibody fragment genes. Therefore, an advanced selection method was required to select from a large number of potential ligands, those capable of recognizing the designated antigen, being the most successful one phage display. Similarly to other *in vitro* selection methods, this technique depends on the capacity to sustain a physical link between a protein and the gene encoding this protein, in this case, between a protein fused to a filamentous phage capsid protein (p3 or p8) presented at the surface of phage M13 and its corresponding gene contained in the encapsidated DNA. After the immunopurification of the molecule by its binding to the selected antigen, the gene is immediately available, leading to sequencing and multiplication of the clone of interest. Because of these *in vitro* selection methods, it is now possible to promptly and efficiently construct fully human antibody fragments against nearly any given antigen [68].

During the same decade, a complementary approach was developed to create fully human antibodies. Transgenic 'humanized' mice were created by replacing the entire mouse IgG repertoire with a human repertoire [68, 69]. During immunization, these humanized mice yield human IgGs and conventional hybridoma techniques are used to clone human antibodies with the desired characteristics. This method has the advantage of producing *in vivo* matured antibodies, avoiding the requirement of further affinity maturation and of directly generate full-length IgG, which is frequently preferred for treatment [68].

In oncology, there is a trend towards the development of 'armed' antibody products, in which the immunoglobulin moiety serves for the selective *in vivo* pharmaco-delivery of bioactive payloads such as cytotoxic drugs, bispecific antibodies, radionuclides or cytokines to sites of disease, thereby sparing healthy tissues.

Indeed, to overcome the fact that most monoclonal antibodies are poor cytocidal agents, monoclonal antibodies are being conjugated to a cytocidal agent, such as toxin or radionuclide, which is then directed to the tumor cell by the antibody. The approach of arming antibodies with toxins has been inspired by the approval of the first immunotoxin-armed antibody - gemtuzumab ozogamicin (Mylotarg), which links the toxin calicheamicin to a CD33-specific antibody and is applied in the treatment of myelogenous leukemia. Furthermore, clinical remissions in patients with hairy cell leukemia are being achieved by treatment with protein-toxin conjugates with a truncated *Pseudomonas* endotoxin genetically linked to CD25- or CD22-specific antibody [51].

Nevertheless, the use of oncotoxins is limited by its immunogenicity that narrows largely the therapeutic window because of the development of antitoxin antibodies and toxicity to normal tissues. Consequently, radiolabeled monoclonal antibodies have been developed as alternative cytotoxic immunoconjugates. Although, before implementing a radioimmunotherapeutic agent regime is obligatory to take into account the monoclonal antibody and the antigenic target, the delivery system used to target the radionuclide to the tumor cell and the radionuclide itself, to assure that most cells attacked are the targeted ones and to reduce radiation exposure to normal tissues [51].

Passive immunotherapy using monoclonal antibodies and receptor Fc-fusion proteins has come of age and has shown great clinical success. The creation of chimeric, humanized or fully human antibodies was a major breakthrough and led to a wave of FDA approved antibodies. Currently, several antibodies are commercialized as therapeutics mainly for cancer and immune disorders [51, 68, 70].

2.5. Passive Immunotherapy: Adoptive T-cell Transfer

There is significant evidence suggesting the efficacy of cell-based therapies, such as, allogeneic transplantation, donor-lymphocyte infusion, and therapies using *ex vivo* expanded T and NK cells, such as lymphokine-activated killer cells, cytokine-induced killer cells, or CD3/CD2843 –expanded cells [57].

Adoptive cell therapy (ACT) consists of a treatment that uses a cancer patient's own T lymphocytes with anti-tumor activity, expanded *in vitro* and reinfused into the patient. This immunotherapeutic option represents the most effective treatment for patients with metastatic melanoma inducing visible cancer regression in approximately 50% of patients. ACT is also associated with clinical improvement in selected patients with post-transplant lymphoproliferative diseases (PTLD) caused by Epstein–Barr virus infection.

Moreover, recent studies have reported that normal lymphocytes can be converted into lymphocytes with anti-cancer activity through genetic modification caused by retroviruses that encode T-cell receptors. These data represented a possibility of extending ACT immunotherapy to patients with a large diversity of cancer types [71].

2.6. Passive Immunotherapy: Immunomodulators

Although they are mainly used in viral diseases, immunomodulators can take part of the therapeutical approach in various other infections namely intracellular bacteria, fungal or parasites. From growth factors to interferons, various drugs are recognized by their immune modulation properties. Therefore, taking into account its increasing use in clinical practice, a detailed presentation of the most common immunomodulators as well as their main indications are further discussed.

2.7. Growth Factors

Growth factors do not have a direct antiviral effect but alter the hematopoiesis, interfering with infections that disrupt it. These substances are usually glycoproteins that alter growth and differentiation of precursors and blood cells [72]. Considering that there are no hematopoietic growth factors licensed for use in veterinary medicine, human growth factors are a common resource. Even though, being a species-specific product, the development of anti-growth factor antibodies occurs, limiting the therapeutics' success [72]. In veterinary medicine, three growth factors are commonly used: filgastrim, sargomastrin and erythropoietin.

2.7.1. Filgastrim

Filgastrim is a granulocyte colony-stimulating factor (G-CSF), commonly used in viral infections in which neutropenias are a severe complication [73, 74]. Only available as a recombinant human product, filgastrim is also used as an adjuvant in chemotherapy protocols, in order to avoid neutropenic side effects [73]. Albeit its clinical potential, this drug may be contraindicated in viral infections once it can lead to an increase of viral replication [72].

2.7.2. Sargramostim

Sargramostim is a granulocyte-macrophage colony-stimulating factor (GM-CSF) which potentiates the replication and differentiation of bone marrow monocyte and granulocyte progenitor cells [75]. Similarly to Filgastrim, this compound is available as a recombinant human product and is commonly administered to increase leucocyte cell counts, particularly in feline retroviral infections such as FIV [76]. It can also be administered in dogs. However, a study reported that high-dose treatment (150µg/kg subcutaneously) induced the production of neutralizing antibodies by this species around 10-12 days after therapy, becoming ineffective and leading to a relapse of leukopenia [77]. Also in cats, the development of neutralizing antibodies occurs in spite of the persistant induced leukocytosis [76]. This suggests that this compound acts differently in this specie. Particularly in FIV-cats, sargramostim is proven to induce a neutrophilia. However, treated cats increased their viral load meaning that this immuno modulator can lead to an increase of viral replication and expression of the virus by peripheral blood mononuclear cells (PBMCs) [72, 76]. Whether it is a good immunomodulator in other infections or not is still unknown. However, due to the neutralizing antibodies and this increase in viral replication, its use must be carefully considered.

2.7.3. Erythropoietin (EPO)

EPO is probably the most commonly used growth factor in small animal practice. Available as a recombinant human product, it is very effective on the treatment of dogs and cats with nonregenerative anemia secondary to endogenous EPO deficiency. Inducing an important stimulation of erythropoiesis, this substance also increases platelet, leukocytes and megakaryocyte cell count [72, 75 - 78]. In opposition to the other previously mentioned growth factors, EPO can be used safely in retroviral infections without inducing an increase on viral load [76].

2.7.4. Insulin-like Growth Factor-1 (IGF-1)

Only available as a recombinant human product, IGF-1 acts as an immuno modulator by stimulating T-lymphocytes. In this sense, it is particularly useful in HIV-infected patients in which T-cell maturation and thymus growth are defected

[72]. In FIV-infected cats, experimental studies have shown that treated cats revealed an elevation of CD4+T cells, an increase in thymus dimensions, a regeneration of thymic cortical and a reduction on B-cell hyperclonal response [79]. Moreover, viral load did not increase meaning that IGF-1 can be a potential immune-modulator for this species. However, taking into account that it should be administered by continuous infusion and it is not easily available for use, IGF-1 is not a common resource in clinical practice.

2.8. Interleukins

Despite their immunomodulation properties, interleukins are rarely used in clinical practice. However, several studies have described and sustained their therapeutic functions in dogs and cats.

2.8.1. Interleukin 2

Being produced by activated helper T-cells, IL-2 stimulates cell-mediated cytotoxicity pathways. In humans, this cytokine can be used in neoplasia or in viral diseases. In fact, IL-2 is reported to induce a more effective immune response to vaccination in HIV infected people [80]. In veterinary medicine, its anti-tumoral properties have been recently described when administered intralesionally in dogs with urinary bladder and uretheral carcinoma as well as non resectable mastocytomas [72].

2.8.2. Interleukin-8

This cytokine has important immunomodulation properties namely in the activation of neutrophils and chemotaxis. Although one study reported its use to potentiate neutrophil activation in dogs [81], there is no data about its therapeutic effect in canine and feline infectious diseases.

2.9. Interferons

Interferons (IFNs) are polypeptides produced by cells of vertebrates after the direct contact with various pathogens such as viruses, bacteria or even molecules such as double chain RNA or bacteria [72]. Being a family of cytokines, they are key components on the activation of the innate immune system being particularly relevant in viral infections [82].

In a molecular point, IFNs can be divided into three classes (I, II and III), taking into account the receptor complex to which they bind [82]. Among these classes type I IFNs are the most studied. In detail, they act as positive feedback inducers, increasing the expression of Toll-like receptors (TLRs) and sensitizing cells to microbial recognition [83]. They establish a truly relation between innate and adaptive immunity, inducing the differentiation of important cells [84]. They are believed not only to block viral replication but also to slow the growth of infected cells making them more susceptible to apoptosis [85, 86].

Due to these referred actions, type I-IFNs are well studied and are frequently used as part of the therapeutic approach in various diseases not only in humans but more recently, in veterinary medicine.

In humans, around 13 subtypes of type I-IFNs are known [82]. Among them, HuIFN-α also called leukocyte interferon is particularly relevant once it can be produced by recombinant DNA technology and is commercially available as an immune modulator drug [72].

In veterinary medicine, the most relevant type I IFN is rFeIFN-ω which was recently developed and is currently licensed for use as an immune modulator not only in cats but also in dogs. Although it is commonly used, in many countries where it is still unavailable, HIFN-α is an immune modulator resource in small animal medicine.

In the further paragraphs, both compounds and the respective therapeutic actions will be discussed.

2.9.1. Human Interferon Alpha (HuIFN- α)

Among their major functions, it was showed that HuIFN-α inhibits oncogenic mutations, it restrains viral nucleic acid synthesis and suppress protein production, being used in various human diseases such as myeloid and hairy cell leukemia, papilomatosis, malignant melanoma, follicular lymphoma, AIDS-related Kaposi sarcoma, hepatitis B and C and HIV [51, 72, 75, 87].

For the same reasons, this compound has been also used in feline medicine namely in FIV, FeLV, Feline Herpesvirus (FHV-1), FCV and Feline Coronavirus (FCoV) viral infections [88 - 92].

Two distinct protocols can be distinguished in feline medicine: the parenteral high dose (10^4-10^6 IU/kg subcutaneously once a day (SID) and the oral low dose (1-50IU/kg SID) [72]. Considering its species-specific potential, when the high dose protocol is administered there is a development of neutralizing antibodies, which occurs within 3-7 weeks after the onset of therapy. Consequently, the immune modulator action of HuIFN-α is limitative and ineffective at a long-.term. In contrast, when HuIFN-α is given orally, it can be administered for a longer period without the arising of neutralizing antibodies. This can be explained by the lower dose administred as well as the route of administration [72, 92].

Concerning retroviral infections, until the release of rFeIFN-ω (further detailed) HuIFN-α were used as a common immunomodulator for FIV and FeLV-infected cats. Nowadays, it is still a valid resource in countries in which rFeIFN-ω is not licensed.

In FIV-infected cats, the recommended high dose protocol was efficient for up to 6-7 weeks, the time at which antibodies developed [72]. Its parenteral use seemed to have a more efficient antiviral effect rather that the oral protocol [93]. The oral protocol (recommended for FIV on the dosage of 50IU/cat SID in cycles of 7 days alternating with 7 days off for 6 months followed by a withdrawal of therapy during 2 months and a repeated cycle of 6 months), is destroyed in the GI tract by gastric acid, trypsin and other proteolytic enzymes [72]. Nonetheless, it improves the clinical status of animals and increases the survival time, which seems to be related to a direct local stimulation of the oral lymphoid tissue [90, 94]. In fact, it seems to directly stimulate the lymphoid local tissue and pharyngeal area, leading to a cascade of immune modulation with up-regulation of IFN-α and a downregulation of IL-4, which can spread for a systemic effect [94].

Also in FeLV, it was shown that this compound inhibits viral replication *in vitro* by decreasing RT activity, increasing apoptosis in infected cells and without affecting viral protein expression [95]. In the high-dose protocol, this compound reduce viral load, although it did not induce a significant clinical improvement [72]. Studies have reported its use in isolated or in combination with antivirals, although in a similar fashion to FIV, the development of neutralizing antibodies limits its therapeutic success [92]. Similarly to what was also referred for FIV and in order to bypass this problem, the oral protocol can be an effective option in FeLV-infected cats [94]. Several authors have reported that this low-dose protocol induces a clinical improvement, believing that it is mainly based on mimic natural defense processes [96].

In sum, despite the fact that higher protocols may have a short-term benefits, the use of oral low dose HuIFN-α protocol is more often performed in feline medicine due to the fact that it can mimicking the immune response towards the non-sel--interferon, having an evident immune-modulator effect.

2.9.2. Recombinant-Feline Interferon Omega (rFeIFN- ω)

As it is called, rFeIFN-ω is a recombinant product produced by a baculovirus which contains the feline IFN sequence. This baculovirus replicates in silkworms after their infection, leading to the production of rFeIFN-ω which is therafter purified [97]. It is currently the only interferon produced for veterinary medicine being licensed for use in Europe, Australia and Japan. *In vitro*, it was shown to have important antiviral, anti-tumoral, antiproliferative and anti-colony effects [98 - 100].

In veterinary practice, it is licensed for use in canine parvovirus and in feline retroviral infections. However, its use has been extended to other viral infections such as FHV-1, FCV or FCoV [101 - 103]. Moreover, its use is also reported in other areas such as oncology [104], dermatology [105] or even endocrinology [106].

In dogs infected with parvovirus, the recommended protocol is 2,5 MU/kg intravenously or subcutaneously for 3 consecutive days [75]. Authors have reinforced it describing that it induces a subtle improvement of clinical signs in dogs infected with parvovirus [107].

Specifically regarding retroviral infections, the recommended protocol consists of

3 cycles of 5 daily subcutaneous injections of 1 MU/kg beginning on days 0, 14 and 60, respectively [108; 75]. Although studies *in vivo* are scarce, it was proven that the referred subcutaneous rFeIFN-ω protocol induces an increase on survival time in FeLV and FIV/FeLV co-infected animals [108]. More recently, other authors evaluated the effect of the licensed protocol in FIV and FeLV infected cats having showed that it does not significantly change viremia, proviral load, Reverse Transcriptase activity or CD4/CD8 ratio [109]. More than improving clinical signs, recent studies have also shown that the licensed rFeIFN protocol decreases concurrent viral excretion in naturally retroviral infected cats, increasing the acute phase proteins profile [110; 111]. Particularly in FIV, in a similar way to HuIFN-alpha, an oral protocol of IFN omega was also successfully described, being a potential alternative in clinical practice [112]. However, clinical improvement seems to be less evident with the oral protocol than with the subcutaneous one. Specifically concerning the immunomodulation properties of both protocols, authors have also studied the effect on Th1/Th2 cytokine profiles describing that, independently of the protocol; this rFeIFN did not significantly change the measured parameters of the acquired immune-response. However, excluding some slightly differences, the production of IL-6 is affected in both protocols sustaining the strong suspicion that IFN-omega acts by reducing pro-inflammatory stimulus (personal data).

Taking all of these findings into account, IFN omega have important immune modulation properties that are extending to areas other than infectious diseases, increasing its utility on the daily routine practice.

An alternative application of T-cell co-stimulatory cytokines includes in-corporation into viral vaccines of their genes. GM-CSF, which acts on dendritic cells, promotes the widest range of T-cell response, including Th1, Th2 and CTL (cytotoxic T lymphocyte). Furthermore, incorporation of the gene encoding IL-12 into DNA vaccines caused antigen-specific predominantly Th1 responses, whereas a Th2 response is induced by inclusion of IL-4 or IL-10 [51].

2.10. Nonspecific Immunostimulants

This "class" of drugs refers to compounds that induce synthesis of IFNs and other cytokines, modulating the immune response. In fact, there are several microorganisms (such as yeasts or intracellular bacteria) that induce the production of various proteins, which have specific antimicrobial properties. For instance, an emulsion obtained from the cell wall of non-pathogenic *Mycobacterium* has been modified and used in equine sarcoid therapy mainly due to its antineoplastic activity. Another example is the use of *Staphylococcus aureus* phage lysate, a sterile preparation based on *Staphylococcus* cell walls with immunostimulant properties, for therapy of pyoderma in dogs (secondary to this bacteria or polymicrobial pyoderma). In fact, these cytokine inducers are particularly attractive in treating infections due to their natural properties of restoring immune function, which allows a better recovery and control of the disease. However, due to the nonspecific immunostimulation that these compounds usually induce, studies *in vitro* cannot be easily extrapolated to *in vivo*, limiting the controlled study designs and impairing an adequate scientific support for these substances [72].

As previously mentioned, in several diseases (particularly if caused by retrovirus) immunomodulation can potentiate the viral replication in once infected macrophages and lymphocytes (namely latently infected ones) usually proliferate with therapy. In this sense, despite the proven clinical benefits, its use should be carefully considered.

2.10.1. Staphylococcal Protein A (SPA)

Produced through the purification of cell walls of *Staphylococcus aureus* Cowan I, this compound has anti-tumoral activity [113, 114]. Among its properties, SPA binds to certain type of immune complexes and stimulates the complement activation as well as natural killer and IFN-gamma production. Due to this immunomodulation functions, several authors have described the use of SPA in experimentally FeLV-infected cats without proven effects on viremia but showing an antitumoral effect [115, 116]. Also in client owned FeLV-infected cats, SPA induced a significant clinical improvement without concurrent changes on survival time or hematologic parameters [117].

2.10.2. Propionibacterium Acnes

Licensed for use in veterinary medicine, *Propionibacterium acnes* results from a killed bacterial product that has important immunomodulator effects such as the increase of natural killer, macrophages and T-cell activity [72, 75]. Interestingly in mice, *Propionibacterium acnes* lead to an increase of NK cells and on the survival time after in-vitro rabies infection [118]. In small animal medicine, its therapeutic use is described in neoplasias (such as malignant melanoma) [119] and as a co-adjuvant in the management of recurrent pyoderma [120]. Also due to its antiviral properties, *Propionibacterium acnes* has been used in feline viral diseases namely FHV, FIP and FeLV [72]. Particularly in FIP, the combination of *P. acnes* and IFN alpha induced an increase on survival time when compared to untreated cats [121]. Although further studies are needed to fully understand its truly immunomodulation effects, *Propionibacterium acnes* seems to be a promisor compound to be used more often in the near future.

2.10.3. Acemannan

Derived from the aloe vera plant, Acemannan is a complex carbohydrate polymer. This compound stimulates the release of cytokines by macrophages, potentiating cell-mediated immunity (namely cytotoxicity response) [72, 75]. Licensed for use in veterinary medicine, Acemannan has been used as a vaccine adjuvant or, intralesionally, on the management of post-vaccinal injection-site sarcomas in cats [122, 123]. In spite of the lack of a controlled study, several authors reported some beneficial effects when it is used (intravenously or subcutaneously) with antibiotics, in a combined supportive therapy with FIV-infected cats [124]. Also in FeLV-cats, it improved survival rate and quality of life [125]. Despite these potential effects, the use of Acemannan as an immunomodulator drug is not frequent in the daily veterinary practice.

2.10.4. Bacille Calmette-Guérin (BCG)

Worldwide known as a vaccine against tuberculosis, BCG refers to cell wall extract of a nonpathogenic strain of *Mycobacterium bovis*. Its use as an immunomodulator drug is mainly based on its properties as a stimulant of the mononuclear-phagocyte clearance mechanisms [75]. In veterinary medicine, old studies have successfully proven its use in the beginnings of small animal oncology, particularly on sticker's sarcoma management [126]. However, its anti-neoplastic effect is questionable once BCG failed to prevent tumor development in cats experimentally inoculated with Feline Sarcoma Virus [127]. Despite these studies, no new ones have been developed making BCG an uncommon immunomodulator in clinical practice.

2.10.5. Serratia Marcescens Extracts

S. marcescens is an anaerobic gram-negative bacillus that is present in soil and water and can cause nosocomial infections (such as urinary or respiratory tract infections) not only in animals but also in humans [128]. From this bacterium, it is possible to produce biologic extracts (containing DNA and membrane components) which have immunomodulation properties. Amongst other functions, these extracts stimulate bone-marrow-derived macrophages to produce pro-inflammatory cytokines (namely IL-1, IL-6 and TNF) that induce fever and an increase on neutrophil's cell count [129]. In FeLV-infected cats, these extracts were shown to induce neutrophilia and fever, although it did not prevent the development of viremia in experimental infected cats [129]. In veterinary oncology, serratia extracts were effectively used on prevention of chemotherapy-induced myelossupression in dogs treated with doxorrubicine [130]. Similarly to BCG, it is not currently used in veterinary practice.

2.10.6. Parapoxvirus

Parapoxvirus (avis or ovis) are inactivated poxviruses licensed for use in veterinary medicine due to their effects on induction of IFNs, activation of NK cells as well as colony-stimulating factors [72, 75]. Despite these functions, studies with control animals are scarce and in the few published double-blind trials with this compound, no beneficial effects were reported.

In FIV and in FIP infections, the use of parapoxvirus is contraindicated since, for the same reasons stated for other immunomodulators, they can lead to a progression of the disease [72]. Curiously in FeLV-infected cats, one study reported that parapoxvirus ovis, when administered subcutaneously one to three times a week (depending on the clinical status of the animal), induced a persistent remission of viremia in a high percentage of animals [131]. However, other authors did not support these findings, questioning whether parapoxvirus compounds are truly efficient [132, 133].

2.11. Drugs From Other Classes That Can Have Immunomodulation Properties

More than their current action, several antibiotics, anti-emetics or antiparasitic drugs have immunomodulation properties. In the further paragraphs, some of these drugs are discussed.

2.11.1. Liposomes

Liposomes are synthetic microscopic structures composed of lipid bilayers surrounded by aqueous layers. These compounds have been used as vehicles to facilitate delivery of various substances, which can be placed within their aqueous compartment [72]. Therefore, they can "transport" multiple drugs, immunologic mediators or even antigens to multiple tissues. By itself, liposomes are notantigenic, not toxic and biodegradable. However, when antigens or immunologic mediators are added to their structure, they are able to stimulate various immune pathways. For instance, liposomal antigens that usually only stimulate humoral responses can activate cell-mediated immunity when specific substances (such as lipid A) are added to their structure.

Due to these properties, liposomes are used as vehicles of several compounds (such as amphotericine) and are particularly relevant in infections by intracellular parasites where the delivery of compounds directly into the mononuclear phagocyte system are particularly useful [72].

2.11.2. Levamisole

Levamisole is an antihelmintic drug derived from the imidazothiazole, very common in veterinary medicine [75, 134]. Among its major properties, this compound modulates the phosphodiesterase activity, leading to an increase of guanosine 3',5'-monophosphate and decrease of adenosine monophosphate. In the immune system, levamisole potentiates cell-mediated immunity by increasing phagocytosis by mononuclear cells, chemotaxis and intracellular destruction of bacteria [72]. Consequently, it has been used as a complementary therapy in various immune-mediated diseases such as peri-anal fistulae. It was also described in feline mammary tumor's therapy but without significant results on survival time [135]. Despite being easily available, toxicity effects (from gastro-enteritis to severe meningoencephalitis) are important and must be carefully considered before its application [136].

2.11.3. Lactoferrin

Derived from mammalian mucosal epithelial cells, lactoferrin is an iron-binding glycoprotein that is present in secretions (such as bovine milk). Being an iron-binding glycoprotein, this compound reduces the iron's availability, having an antimicrobial effect [72, 75]. Moreover, lactoferrin increases mononuclear cells phagocytic activity in humans, having documented protective effects during lethal bacteriemia in mice [137]. Also in FIV-infected cats, this product has shown to improve clinical signs of stomatitis, increasing the phagocytic activity of neutrophils [138]. In clinical practice, lactoferrin seems to be a useful compound for topical use, particularly in cases of severe oral stomatitis [72].

2.11.4. Nosodes

These compounds are homeopathic preparations made from infected tissues of patients that had the disease to which Nosodes are intended to prevent. Although the use of homeopathic compounds has evoked several problems (mainly due to its uncontrolled policy) [139], it is believed that nosodes has a specific immunomodulatory effect also in viral infections of cats and dogs [72]. However, only one controlled study documented its administration in puppies infected by canine parvovirus which was ineffective [72, 140]. The lack of clinical trials in veterinary medicine impairs a more confident use of this product.

2.11.5. Cimetidine

Cimetidine is a reversible H2-receptor antagonist that, by competitive binding to histamine receptors, reduces gastric acidity. Therefore, it is widely used in clinical practice. Besides these indications, it is believed that cimetidine potentiates cell-mediated immunity in humans by blocking receptors on suppressor T-cells.

Although there are no controlled studies in dogs and cats, this compound seems to be an indirect immunomodulator [72].

In summary, independently of their main function, several immunomodulators are currently available for use in veterinary medicine. In spite of the various substances extrapolated from the human medicine, the lack of control studies impairs a more frequent use. Nevertheless, immunomodulator compounds seem to be a useful therapeutic tool in the near future, contributing to reduce antibiotic therapy and inducing a more focal approach of animal infections.

2.11.6. Active Immunotherapy

In 1891, William Coley, a young surgeon, performed intratumoral injections of live or inactivated *Streptococcus pyogenes* and *Serratia marcescens* with the intention of reproducing the spontaneous remissions of sarcomas observed in rare-cancer patients who had developed erysipelas. Given Elie Metchnikoff's contemporaneous work demonstrating the immune system's ability to cause inflammation and destroy invading bacteria, 'Coley's toxins' activity could be explained by stimulation and activation of antibacterial phagocytes that would kill bystander tumor cells [141]

Although vaccines have been quite successful for preventing infectious disease, imunnotherapies for established chronic infectious diseases or cancer, have been much less effective. Nevertheless, recent immunological insights motivated progresses toward this challenge [51].

2.12. Active Immunotherapy: Vaccines

In April ,2010 the first cancer vaccine was approved for clinical use in patients with advanced prostate cancer after placebo-controlled trials showed an improved overall survival [142]. Since then, a wide spectrum of other cancer vaccine platforms that target a diverse range of tumor-associated antigens is currently being evaluated in randomized phase II and phase III trials, including injection of peptides or proteins in adjuvant, injection of recombinant viruses or other recombinant microorganisms, delivery of killed tumor cells, or delivery of protein- or peptide-activated dendritic cells (DCs) to the patient [143].

Cancer vaccines represent a highly attractive and innovative approach to cancer therapy because they direct and stimulate the patient's own immune system against cancer. Cancer vaccines educate the immune system to respond to tumor-associated antigens (TAAs), inducing the recognition, attack and destruction of tumors [51, 143, 144].

Each of the platforms in development has advantages and disadvantages that are determined by the specific tumor-associated antigen (TAA) targeted, the disease and disease stage, the clinical trial endpoint, and whether the vaccine is assessed in combination with an immune stimulant, an immune suppression inhibitor, or another approach of cancer therapy [143].

The vaccine target viability will be influenced by the capacity of T-cells or B-cells to recognize the TAA in the context of MHC, the level of expression, the tumor specificity of the TAA compared with normal adult tissue, and the extent of intrinsic “tolerance” to the particular TAA [143, 145].

In fact, a major challenge for the development of effective vaccines is the identification of TAA that stimulate T-cell responses. These include tumor-specific antigens, the results of mutations, viral antigens in cancers induced by viruses, and tumor-specific differentiation antigens. Viruses etiologically associated with neoplasias, including hepatitis B, hepatitis C (hepatoma), human papilloma viruses (cervical cancer), Epstein-Barr virus (Burkitt lymphoma, Hodgkin disease) and human T-cell lymphotrophic virus-1 (T-cell leukemia and lymphoma) appear to promote host immune responses. These agents are currently under study or being employed in vaccines [51].

The immunoglobulin idiotype (Id) characterizes an alternative antigenic target that occurs in B-cell lymphomas [51]. As clonal products of a particular cell progenitor, each lymphoma cell expresses a specific Id, different from every other B cell, and consequently representing a unique TAA. Regarding that passive immunotherapy with monoclonal antibodies targeting B-cell lymphoma Ids induces significant clinical responses; there is an expectation that active vaccines that promote a polyclonal antibody response would be similarly effective. Although TAA in its natural form might be poorly immunogenic, it’s immunogenicity can be improved by chemical conjugation with a carrier protein, like keyhole limpet hemocyanin (KLH), and by combining it with an immune-stimulant, such as granulocyte-macrophage colony-stimulating factor (GM-CSF). A limitation of this approach is that, unless the targeted TAA is shared between tumors from different patients, each anti-idiotype vaccine has to be developed specifically for each patient and therefore involving several resources for its discovery and production [51, 57, 143].

Another technique used to define cancer-associated antigens is to study antigens recognized by the host. For instance, the serological identification by recombinant expression cloning (SEREX) is a method that analyses circulating IgG that recognize TAAs. Furthermore, an alternative approach that could induce anti-tumor cellular immunity rather than humoral immune responses is to screen cDNA libraries extracted from tumor tissues using tumor-reactive T-cell lines and clones obtained from cancer bearing patients. Finally, other method is to investigate tumor-associated peptides linked to class I major histocompatibility (MHC) molecules by mass spectrometry. Likewise, melanocyte-specific antigens, including MAGE, MART-1/melanA, tyrosinase, tyrosine-related protein 1, and gp100, were correlated with antitumoral clinical responses and with vitiligo subsequent to immunological attack of either malignant and normal melanocytes. The current advances on cancer molecular profiling field may contribute for the discovery of additional valuable antigenic targets suitable for immunotherapy [51].

A number of different cancer vaccine platforms are under active clinical development, including recent insights into tumor rejection antigens. These research projects may be based on molecular biology, directed to specific antigens that are provided as peptide, protein, or as genetically engineered plasmid DNA vectors, bacteria, or viruses. Alternatively, they may be based on cell biology, using patient-derived dendritic cells, autologous tumor cells or tumor cell lysates, or allogeneic tumor cells [144].

2.13. Peptide Plus Adjuvant

Vaccines developed with peptides from TAAs are considered one of the simplest formulations and are generally delivered with an adjuvant and/or with an immunomodulator. This vaccine platform has a number of advantages over other types of cancer vaccines. On one hand, it allows to evaluate the patients' immune responses since its development requires the exact knowledge of relevant epitopes, and, on the other hand, its manufacturing on a large scale is cost-effective. However, drawbacks include poor immunogenicity, since they target only one epitope or a few epitopes of the TAA and the requirement of an appropriate patient's MHC type. It is considered that a cancer vaccine optimal efficacy relies on antigen-specific CD8+ cytolytic T cells (CTLs) response, which is responsible for tumor cell lysis, and on antigen-specific CD4+ "helper" T cells response. Protein-based vaccines are more expensive than peptide-based vaccines; however they generally present CD4 and CD8 interacting epitopes. Several peptide-based and protein-based vaccines are also being developed as part of a Dendritic cells (DCs) vaccine [143 - 145].

2.14. Plasmid DNA

Another relatively simple platform for cancer vaccine formulation consists of plasmid DNA. Similarly to peptide vaccines it requires adequate knowledge of important target antigens in order to develop the plasmid vector containing the correct gene sequences.

Plasmid DNA vaccines also are regarded as easy and inexpensive to produce, relatively safe and easy to distribute in areas with limited resources such as underdeveloped countries. However, their weak immunogenicity *in vivo* has been responsible for a restricted activity. In order to improve immunogenicity, these plasmid DNA vaccines can be combined with adjuvants, or more advanced delivery approaches can be used, such as gene gun techniques or electroporation. As an alternative, the plasmid DNA vector might be engineered with regulatory sequences to express high levels of the desired gene product [144]. Although DNA vaccine platforms have shown to be promising in preclinical studies, early clinical trials so far have been disappointing [143, 145].

2.15. Recombinant Virus

Recombinant viral vectors have been used in the development of cancer immunotherapy approaches for the past 10 years and represent an alternative strategy to deliver specific tumor antigens *via* gene transfer. These vectors include replication-defective pox viruses, adenoviruses, adeno-associated viruses, herpesviruses, and retroviral or lentiviral vectors. Unlike plasmid DNA vectors, the immunogenic potential of such vectors is high because they are self-adjuvanted and offer the ability to express multiple tumor-associated antigens (TAAs) along with an array of immune co-factors. Though, the potential for toxicity is also high, and is associated with the risk of an active infection, recombination with other infectious viruses, or endogenous retroviruses. The regulation of gene transfer approach remains difficult, resulting in a problematic manufacturing and commercial development process [144, 146, 147].

2.16. Recombinant Bacteria

Recombinant bacteria represent a fourth strategy to deliver defined tumor antigens by gene transfer. The antigen delivery platforms include *Listeria monocytogenes*, *Salmonella Typhimurium*, *Shigella*spp , and *Mycobacterium bovis* BCG. The immunogenic potential as well as the possibility of toxicity with this vaccination approach is also rather significant. These gene transfer platforms remain in early-phase clinical trials [144].

2.17. Dendritic Cell Vaccines

Both immunity and tolerance are controlled by a network of professional antigen-presenting cells (APCs). Dendritic cells are professional APC, and provide the most potent stimulation of naive T cells. An approach gaining increasing acceptance in the tumor vaccine field is to immunize cancer patients with their own DCs loaded *ex vivo* with tumor antigens, including peptide-based or protein-based, or viral vector–infected, DCs [144, 146, 148, 149].

The Sipuleucel-T vaccine, which was recently approved by the US Food and Drug Administration (FDA) for the therapy of asymptomatic metastatic castrate-resistant prostate cancer (mCRPC), is designed to stimulate an immune response to prostatic acid phosphatase (PAP), an antigen that is found on most prostate cancer cells. Unlike some other cancer treatment vaccines under development [150], Sipuleucel-T vaccine is customized to each patient and consists of APCs from peripheral blood mononuclear cells (PBMCs) that have been incubated with prostatic acid phosphatase (PAP, a prostate antigen) fused to granulocyte macrophage colony-stimulating factor (GM-CSF). This vaccine regimen involves the isolation of APCs from the patient through leukaphereses. The APCs are sent to a central facility where they are cultured with a protein called PAP-GM-CSF. This protein consists of PAP linked to another protein called granulocyte-macrophage colony-stimulating factor (GM-CSF). These cells are then administered to the patient; this procedure is repeated three times at biweekly intervals. One disadvantage of DC and/or APC platforms is that they involve leukapheresis and cell culture processing of PBMCs, and therefore only a restricted number of vaccinations can be performed [143].

2.18. Tumor Cell Vaccines

The use of whole tumor–cell vaccines has the advantage of presenting a wide range of both known and indeterminate TAAs, leaving it to the immune system to choose its specific protein targets. Nevertheless, this characteristic also potentially reduces the relative level of expression of a specific TAA or group of TAAs and consequently its presentation and processing by APCs. Due to the fact that tumor cells are not innately immunogenic, their immunogenicity usually must be enhanced by genetic, viral, or chemical modification, or by the co-delivery of a strong immune stimulant such as GM-CSF, Bacillus Calmette–Guerin adjuvant (BCG), or CD40 ligand (CD40L) [143, 144].

As referred, autologous tumor cell vaccines have the important asset of presenting an exclusive set of TAAs, such as specific point mutations or fusion gene products, from a patient's own tumor. Therefore, this strategy relies on the possibility of performing tumor biopsies and this is only viable for some tumor types and stages. As a variant of this method, before immunization of the patient, DCs and autologous tumor cells are fused, combining the unique particularities of whole tumor–cell vaccines with the improved antigen-presenting ability of DCs [144, 145].

Alternatively, allogeneic whole tumor–cell vaccines, generally comprising two or three well established and characterized human tumor cell lines of a particular tumor type, may overcome several logistical limitations related to autologous tumor–cell vaccines [144 - 146]. The GVAX vaccine platforms, which include allogeneic pancreatic, prostate, or breast tumor cells, show evidences of the ability to offer such a vaccine for multicenter assessment [144, 145].

2.19. Heat-shock Protein and Exosome-based Vaccines

Heat-shock proteins (HSPs) resultant from tumor cells, including gp96, hsp90, hsp70, calreticulin, hsp110, and hsp170, present strong immunogenicity. The chaperone function of HSPs allows their combination with immunogenic tumor specific peptides [144].

Exosome-based vaccines are another platform that promotes tumor immunity. An exosome is a non-plasma membrane-derived vesicle formed through the fusion of internal vesicles with the plasma membrane, and released into the external environment as exosomes.

The presence of antigens on exosomes released from tumor cells and their capacity to induce anti-tumor responses *in vitro* and *in vivo* suggest that this vaccine platform may hold therapeutic promise for cancer patients [151].

Exosomes can be obtained directly from DCs (DEX), inducing an immune response by the *in vitro* transfer of MHC:peptide complexes to naive DC. Remarkably, it is possible to improve the immunogenicity of DEX by developing them from DC matured by lipopolysaccharide rather than from DC that were not subjected to a maturation process. As an alternative, exosomes can be derived directly from tumor cells (TEX), which comprise the antigenic set (MHC Class I, HSP, and tumor antigens). However, TEX have the potential to cause an immune tolerance, consequently, they should be pulsed onto DC matured *ex vivo*, or be derived from heat-activated tumor cells to promote tumor immunity rather than inhibit it [144].

2.20. Active Immunotherapy: Checkpoint Blockade

There are several obstacles to the effective immunotherapy of cancer that may limit the success of treatment of cancer patients with minimal disease or may promote merely transient tumor responses. Some of these complications are tumor cell related, though the major result from a series of negative immunoregulatory elements (*i.e.* Tregs, myeloid-derived suppressor cells, tumor-associated macrophages, vascular endothelial growth factor, TGFβ, IL-10) that are normally responsible for the prevention of self-reactive destructive immune responses that lead to autoimmune disease [51, 152, 153].

These immunoregulatory mechanisms correlate with poor clinical success in cancer patients and limit the outcome of vaccines in preclinical models. Therefore, a focus of recent preclinical tumor immunology has been to study various types of checkpoint blockade, either to improve vaccine-induced immunity or to reveal an underlying antitumor immune response. One approach is the use of mAbs directed to these checkpoints, and there is considerable preclinical evidence demonstrating the effect of agonistic antibodies specific for 4-1BB and OX4 mAbs or antagonistic mAbs specific for coinhibitory receptors such as CTLA4 or PD-1 [51 - 57, 146 - 152].

Effective immunotherapy for chronic infectious diseases or cancer will require the use of appropriate target antigens, the optimization of the interaction between the antigenic peptide, the antigen presenting cell (APC) and the T cell, and the simultaneous blockade of negative regulatory mechanisms that hamper immuno-therapy effects [51].

CONCLUDING REMARKS

Multi-resistant infections are a major threat and continue to demand urgent alternatives to antibiotics. Immunotherapy appears to be an excellent strategy in finding solutions to overcome antibiotic resistance. Its applications extend to the treatment of not only multidrug resistant bacterial infections, but also autoimmune and oncological diseases. Its potential has led many scientists to consider it the most significant scientific advance of the last 20 years. Let's wait for developments over the next 20!

CONFLICT OF INTEREST

The authors confirm that this chapter contents have no conflict of interest.

ACKNOWLEDGEMENT

CIISA/FMV for financial support of two projects: "Clinical and immunological characterization of naturally occurring canine lymphoma: Development and application of engineered recombinant antibodies for diagnosis and treatment". (CIISA 88 Linfoma Canino) and "Feline Interferon Omega as a therapeutic agent in cats naturally infected with Feline Immunodeficiency virus (FIV)" (CIISA50 Viroses Felinas).

REFERENCES

[1] Parham P. The Immune System. (3rd.), New york: Garland Science, Taylor & Francis Group 2009.

[2] Peterson J. Bacterial Pathogenesis In: Baron S, Ed. Medical Microbiology. (4th.), Galveston: University of Texas Medical Branch at Galveston 1996.

[3] Aggarwal A. Medical Microbiology. (1st.), New Delhi: International Scientific Publishing Academy 2010.

[4] Wiesner J, Vilcinskas A. Antimicrobial peptides: the ancient arm of the human immune system. Virulence 2010; 1(5): 440-64. [http://dx.doi.org/10.4161/viru.1.5.12983] [PMID: 21178486]

[5] Bardan A, Nizet V, Gallo RL. Antimicrobial peptides and the skin. Expert Opin Biol Ther 2004; 4(4): 543-9. [http://dx.doi.org/10.1517/14712598.4.4.543] [PMID: 15102603]

[6] Schauber J, Gallo RL. Antimicrobial peptides and the skin immune defense system. J Allergy Clin Immunol 2009; 124(3) (Suppl. 2): R13-8. [http://dx.doi.org/10.1016/j.jaci.2009.07.014] [PMID: 19720207]

[7] Bernard JJ, Gallo RL. Protecting the boundary: the sentinel role of host defense peptides in the skin. Cell Mol Life Sci 2011; 68(13): 2189-99. [http://dx.doi.org/10.1007/s00018-011-0712-8] [PMID: 21573782]

[8] Steinstraesser L, Kraneburg U, Jacobsen F, Al-Benna S. Host defense peptides and their antimicrobial-immunomodulatory duality 322-3. [http://dx.doi.org/10.1016/j.imbio.2010.07.003]

[9] Guaní-Guerra E, Santos-Mendoza T, Lugo-Reyes SO, Terán LM. Antimicrobial peptides: general overview and clinical implications in human health and disease. Clin Immunol 2010; 135(1): 1-11. [http://dx.doi.org/10.1016/j.clim.2009.12.004] [PMID: 20116332]

[10] Nakamizo S, Egawa G, Honda T, *et al.* Commensal bacteria and cutaneous immunity. Semin Immunopathol 2014. [PMID: 25326105]

[11] Weese JS. The canine and feline skin microbiome in health and disease. Vet Dermatol 2013; 137-45. [http://dx.doi.org/10.1002/9781118644317.ch19]

[12] Schommer NN, Gallo RL. Structure and function of the human skin microbiome. Trends Microbiol 2013; 21(12): 660-8.
[http://dx.doi.org/10.1016/j.tim.2013.10.001] [PMID: 24238601]

[13] Schloss PD. Microbiology: An integrated view of the skin microbiome. Nature 2014; 514(7520): 44-5.
[http://dx.doi.org/10.1038/514044a] [PMID: 25279916]

[14] Grice EA, Segre JA. The skin microbiome. Nat Rev Microbiol 2011; 9(4): 244-53.
[http://dx.doi.org/10.1038/nrmicro2537] [PMID: 21407241]

[15] Zhang M, Jiang Z, Li D, Jiang D, *et al.* Oral antibiotic treatment induces skin microbiota dysbiosis and influences wound healing. Microb Ecol 2014.
[PMID: 25301498]

[16] Sanford JA, Gallo RL. Functions of the skin microbiota in health and disease. Semin Immunol 2013; 25(5): 370-7.
[http://dx.doi.org/10.1016/j.smim.2013.09.005] [PMID: 24268438]

[17] Manring MM, Hawk A, Calhoun JH, Andersen RC. Treatment of war wounds: a historical review. Clin Orthop Relat Res 2009; 467(8): 2168-91.
[http://dx.doi.org/10.1007/s11999-009-0738-5] [PMID: 19219516]

[18] Wilson B, Salyers A, Whitt D, Winkler M. Bacterial pathogenesis-A molecular approach. 2011.
[http://dx.doi.org/10.1128/9781555816162]

[19] Vitetta L, Manuel R, Zhou JY, Linnane AW, Hall S, Coulson S. The overarching influence of the gut microbiome on end-organ function: the role of live probiotic cultures. Pharmaceuticals (Basel) 2014; 7(9): 954-89.
[http://dx.doi.org/10.3390/ph7090954] [PMID: 25244509]

[20] Taschuk R, Griebel PJ. Commensal microbiome effects on mucosal immune system development in the ruminant gastrointestinal tract. Anim Health Res Rev 2012; 13(1): 129-41.
[http://dx.doi.org/10.1017/S1466252312000096] [PMID: 22853940]

[21] Chung H, Kasper DL. Microbiota-stimulated immune mechanisms to maintain gut homeostasis. Curr Opin Immunol 2010; 22(4): 455-60.
[http://dx.doi.org/10.1016/j.coi.2010.06.008] [PMID: 20656465]

[22] Tlaskalova-Hogenova H, Stepankova R, Hudcovic T, Tuckova L, Cukrowska B, Lodinova-Zadnikova R, *et al.* et al Commensal bacteria (normal microflora), mucosal immunity and chronic inflammatory and autoimmune diseases Immunol Lett 2004; (2-3): 97-108.

[23] Guinane CM, Cotter PD. Role of the gut microbiota in health and chronic gastrointestinal disease: understanding a hidden metabolic organ. Therap Adv Gastroenterol 2013; 6(4): 295-308.
[http://dx.doi.org/10.1177/1756283X13482996] [PMID: 23814609]

[24] Qin J, Li R, Raes J, *et al.* A human gut microbial gene catalogue established by metagenomic sequencing. Nature 2010; 464(7285): 59-65.
[http://dx.doi.org/10.1038/nature08821] [PMID: 20203603]

[25] Ostaff MJ, Stange EF, Wehkamp J. Antimicrobial peptides and gut microbiota in homeostasis and pathology. EMBO Mol Med 2013; 5(10): 1465-83.
[http://dx.doi.org/10.1002/emmm.201201773] [PMID: 24039130]

[26] Mantis NJ, Rol N, Corthésy B. Secretory IgA's complex roles in immunity and mucosal homeostasis in the gut. Mucosal Immunol 2011; 4(6): 603-11.
[http://dx.doi.org/10.1038/mi.2011.41] [PMID: 21975936]

[27] Nishifuji K, Sugai M, Amagai M. Staphylococcal exfoliative toxins: "molecular scissors" of bacteria that attack the cutaneous defense barrier in mammals. J Dermatol Sci 2008; 49(1): 21-31.
[http://dx.doi.org/10.1016/j.jdermsci.2007.05.007] [PMID: 17582744]

[28] McGuckin MA, Lindén SK, Sutton P, Florin TH. Mucin dynamics and enteric pathogens. Nat Rev Microbiol 2011; 9(4): 265-78.
[http://dx.doi.org/10.1038/nrmicro2538] [PMID: 21407243]

[29] Diebel LN, Liberati DM, Baylor AE, Brown WJ, Devlin J. Immunoglobulin A protease is a virulence factor for gram-negative pneumonia. Surgery 2004; 136(4): 937-43.
[http://dx.doi.org/10.1016/j.surg.2004.06.029] [PMID: 15467681]

[30] Kraus D, Peschel A. Molecular mechanisms of bacterial resistance to antimicrobial peptides. Curr Top Microbiol Immunol 2006; 306: 231-50.
[http://dx.doi.org/10.1007/3-540-29916-5_9] [PMID: 16909924]

[31] Peschel A. How do bacteria resist human antimicrobial peptides? Trends Microbiol 2002; 10(4): 179-86.
[http://dx.doi.org/10.1016/S0966-842X(02)02333-8] [PMID: 11912025]

[32] Nizet V. Antimicrobial peptide resistance mechanisms of human bacterial pathogens. Curr Issues Mol Biol 2006; 8(1): 11-26.
[PMID: 16450883]

[33] Tilly K, Rosa PA, Stewart PE. Biology of infection with Borrelia burgdorferi. Infect Dis Clin North Am 2008; 22(2): 217-34.
[http://dx.doi.org/10.1016/j.idc.2007.12.013] [PMID: 18452798]

[34] Caza M, Kronstad JW. Shared and distinct mechanisms of iron acquisition by bacterial and fungal pathogens of humans. Front Cell Infect Microbiol 2013; 3: 80.
[http://dx.doi.org/10.3389/fcimb.2013.00080] [PMID: 24312900]

[35] Friedman DB, Stauff DL, Pishchany G, Whitwell CW, Torres VJ, Skaar EP. Staphylococcus aureus redirects central metabolism to increase iron availability. PLoS Pathog 2006; 2(8): e87.
[http://dx.doi.org/10.1371/journal.ppat.0020087] [PMID: 16933993]

[36] Kwinn LA, Nizet V. How group A Streptococcus circumvents host phagocyte defenses. Future Microbiol 2007; 2(1): 75-84.
[http://dx.doi.org/10.2217/17460913.2.1.75] [PMID: 17661677]

[37] Clarke SC. Bacteria as potential tools in bioterrorism, with an emphasis on bacterial toxins. Br J Biomed Sci 2005; 62(1): 40-6.
[PMID: 15816214]

[38] Anderson PD. Bioterrorism: toxins as weapons. J Pharm Pract 2012; 25(2): 121-9.
[http://dx.doi.org/10.1177/0897190012442351] [PMID: 22523138]

[39] Henkel JS, Baldwin MR, Barbieri JT. Toxins from bacteria. EXS 2010; 100: 1-29.
[http://dx.doi.org/10.1007/978-3-7643-8338-1_1] [PMID: 20358680]

[40] Nikaido H. Multidrug resistance in bacteria. Annu Rev Biochem 2009; 78: 119-46. [http://dx.doi.org/10.1146/annurev.biochem.78.082907.145923] [PMID: 19231985]

[41] Antimicrobial resistance. Fact sheet n°184: World Health Organizationupdated 2014 April;

[42] Moodley A, Damborg P, Nielsen SS. Antimicrobial resistance in methicillin susceptible and methicillin resistant *Staphylococcus pseudintermedius* of canine origin: literature review from 1980 to 2013. Vet Microbiol 2014; 171(3-4): 337-41. [http://dx.doi.org/10.1016/j.vetmic.2014.02.008] [PMID: 24613081]

[43] Braga ED, Aguiar-Alves F, de Freitas MdeF, *et al.* High prevalence of *Staphylococcus aureus* and methicillin-resistant S. aureus colonization among healthy children attending public daycare centers in informal settlements in a large urban center in Brazil. BMC Infect Dis 2014; 14: 538. [http://dx.doi.org/10.1186/1471-2334-14-538] [PMID: 25287855]

[44] Ishihara K, Saito M, Shimokubo N, Muramatsu Y, Maetani S, Tamura Y. Epidemiological analysis of methicillin-resistant *Staphylococcus aureus* carriage among veterinary staff of companion animals in Japan. J Vet Med Sci 2014; 76(12): 1627-9. [http://dx.doi.org/10.1292/jvms.14-0119] [PMID: 25649946]

[45] Mehraj J, Akmatov MK, Strömpl J, *et al.* Methicillin-sensitive and methicillin-resistant Staphylococcus aureus nasal carriage in a random sample of non-hospitalized adult population in northern Germany. PLoS One 2014; 9(9): e107937. [http://dx.doi.org/10.1371/journal.pone.0107937] [PMID: 25251407]

[46] Yeaman MR, Filler SG, Schmidt CS, Ibrahim AS, Edwards JE Jr, Hennessey JP Jr. Applying convergent immunity to innovative vaccines targeting *Staphylococcus aureus*. Front Immunol 2014; 5: 463. [http://dx.doi.org/10.3389/fimmu.2014.00463] [PMID: 25309545]

[47] Bagnoli F, Bertholet S, Grandi G. Inferring reasons for the failure of *Staphylococcus aureus* vaccines in clinical trials. Front Cell Infect Microbiol 2012; 2: 16. [http://dx.doi.org/10.3389/fcimb.2012.00016] [PMID: 22919608]

[48] Joice R, Lipsitch M. Targeting imperfect vaccines against drug-resistance determinants: a strategy for countering the rise of drug resistance. PLoS One 2013; 8(7): e68940. [http://dx.doi.org/10.1371/journal.pone.0068940] [PMID: 23935910]

[49] Karauzum H, Adhikari RP, Sarwar J, *et al.* Structurally designed attenuated subunit vaccines for S. aureus LukS-PV and LukF-PV confer protection in a mouse bacteremia model. PLoS One 2013; 8(6): e65384. [http://dx.doi.org/10.1371/journal.pone.0065384] [PMID: 23762356]

[50] Hamblin T, Ed. Immunotherapy of Disease. Kluwer Academic 1990. [http://dx.doi.org/10.1007/978-94-009-1844-3]

[51] Waldmann TA. Immunotherapy: past, present and future. Nat Med 2003; 9(3): 269-77. [http://dx.doi.org/10.1038/nm0303-269] [PMID: 12612576]

[52] Nara PL. Humoral immunity. In: Jacobson J, Ed. Immunotherapy for Infectious Diseases. United States of America: Humana Press 2002. [http://dx.doi.org/10.1385/1-59259-171-X:03]

[53] Stern AM, Markel H. The history of vaccines and immunization: familiar patterns, new challenges. Health Aff (Millwood) 2005; 24(3): 611-21. [http://dx.doi.org/10.1377/hlthaff.24.3.611] [PMID: 15886151]

[54] Akdis M. New treatments for allergen immunotherapy. World Allergy Organ J 2014; 7(1): 23. [http://dx.doi.org/10.1186/1939-4551-7-23] [PMID: 25258656]

[55] Helbig ET, Opitz B, Sander LE. Adjuvant immunotherapies as a novel approach to bacterial infections. Immunotherapy 2013; 5(4): 365-81. [http://dx.doi.org/10.2217/imt.13.17] [PMID: 23557420]

[56] Feldmann M, Steinman L. Design of effective immunotherapy for human autoimmunity. Nature 2005; 435(7042): 612-9. [http://dx.doi.org/10.1038/nature03727] [PMID: 15931214]

[57] Brody J, Kohrt H, Marabelle A, Levy R. Active and passive immunotherapy for lymphoma: proving principles and improving results. J Clin Oncol 2011; 29(14): 1864-75. [http://dx.doi.org/10.1200/JCO.2010.33.4623] [PMID: 21482977]

[58] Lizée G, Overwijk WW, Radvanyi L, Gao J, Sharma P, Hwu P. Harnessing the power of the immune system to target cancer. Annu Rev Med 2013; 64: 71-90. [http://dx.doi.org/10.1146/annurev-med-112311-083918] [PMID: 23092383]

[59] Malki A, ElSharkawy A. Chapter: Tumor Immunology and Immunotherapy In: Malki A, Eds Cancer Treatment Strategies, USA:OMICS Group eBooks Chapter. 2014.

[60] Guo JT, Bichko VV, Seeger C. Effect of alpha interferon on the hepatitis C virus replicon. J Virol 2001; 75(18): 8516-23. [http://dx.doi.org/10.1128/JVI.75.18.8516-8523.2001] [PMID: 11507197]

[61] Marrack P, Kappler J, Kotzin BL. Autoimmune disease: why and where it occurs. Nat Med 2001; 7(8): 899-905. [http://dx.doi.org/10.1038/90935] [PMID: 11479621]

[62] Chandrashekara S. The treatment strategies of autoimmune disease may need a different approach from conventional protocol: a review. Indian J Pharmacol 2012; 44(6): 665-71. [http://dx.doi.org/10.4103/0253-7613.103235] [PMID: 23248391]

[63] Drake C. Combination immunotherapy approaches, Ann Oncol. Suppl 2012; 8(8): 41-6.

[64] Dimitrov DS, Marks JD. Therapeutic antibodies: current state and future trends--is a paradigm change coming soon? Methods Mol Biol 2009; 525: 1-27, xiii. [http://dx.doi.org/10.1007/978-1-59745-554-1_1] [PMID: 19252861]

[65] Yamada T. Therapeutic monoclonal antibodies. Keio J Med 2011; 60(2): 37-46. [http://dx.doi.org/10.2302/kjm.60.37] [PMID: 21720199]

[66] Norman DJ. Mechanisms of action and overview of OKT3. Ther Drug Monit 1995; 17(6): 615-20. [http://dx.doi.org/10.1097/00007691-199512000-00012] [PMID: 8588230]

[67] LoBuglio AF, Wheeler RH, Trang J, *et al.* Mouse/human chimeric monoclonal antibody in man: kinetics and immune response. Proc Natl Acad Sci USA 1989; 86(11): 4220-4. [http://dx.doi.org/10.1073/pnas.86.11.4220] [PMID: 2726771]

[68] Chames P, Van Regenmortel M, Weiss E, Baty D. Therapeutic antibodies: successes, limitations and hopes for the future. Br J Pharmacol 2009; 157(2): 220-33.
[http://dx.doi.org/10.1111/j.1476-5381.2009.00190.x] [PMID: 19459844]

[69] Lonberg N. Human monoclonal antibodies from transgenic mice. Handbook Exp Pharmacol 2008; 181(181): 69-97.
[http://dx.doi.org/10.1007/978-3-540-73259-4_4] [PMID: 18071942]

[70] Nelson AL, Dhimolea E, Reichert JM. Development trends for human monoclonal antibody therapeutics. Nat Rev Drug Discov 2010; 9(10): 767-74.
[http://dx.doi.org/10.1038/nrd3229] [PMID: 20811384]

[71] Rosenberg SA, Restifo NP, Yang JC, Morgan RA, Dudley ME. Adoptive cell transfer: a clinical path to effective cancer immunotherapy. Nat Rev Cancer 2008; 8(4): 299-308.
[http://dx.doi.org/10.1038/nrc2355] [PMID: 18354418]

[72] Hartmann K. Chapter 2: Antiviral and Immunomodulatory Chemotherapy. In: Greene C, Ed. Infectious Diseases of the Dog and Cat. (4th.) United States of America: Elsevier Saunders 2012; pp. 10-24.

[73] Kraft W, Kuffer M. [Treatment of severe neutropenias in dogs and cats with filgrastim]. Tierarztl Prax 1995; 23(6): 609-13.
[PMID: 8585081]

[74] Rewerts JM, McCaw DL, Cohn LA, Wagner-Mann C, Harrington D. Recombinant human granulocyte colony-stimulating factor for treatment of puppies with neutropenia secondary to canine parvovirus infection. J Am Vet Med Assoc 1998; 213(7): 991-2.
[PMID: 9776994]

[75] Greene C, Calpin J. Appendix: Antimicrobial drug formulary Em C Greene, Infectious Diseases of the Dog and Cat. (4th ed.) Elsevier Saunders 2012; pp. 1207-320.

[76]]Arai M, Darman J, Lewis A. et al. The use of human hematopoietic growth factors (rhGM-CSF and rhEPO) as a supportive therapy for FIV-infected cats Vet Immunol Immunop 2000; 77(1-2): 71-92.

[77] Mayer P, Werner FJ, Lam C, Besemer J. *In vitro* and *in vivo* activity of human recombinant granulocyte-macrophage colony-stimulating factor in dogs. Exp Hematol 1990; 18(9): 1026-33. [PMID: 2168838]

[78] Ogilvie GK. Hematopoietic growth factors: frontiers for cure. Vet Clin North Am Small Anim Pract 1995; 25(6): 1441-56.
[http://dx.doi.org/10.1016/S0195-5616(95)50163-9] [PMID: 8619275]

[79] Woo JC, Dean GA, Lavoy A, Clark R, Moore PF. Investigation of recombinant human insulin-like growth factor type I in thymus regeneration in the acute stage of experimental FIV infection in juvenile cats. AIDS Res Hum Retroviruses 1999; 15(15): 1377-88.
[http://dx.doi.org/10.1089/088922299310089] [PMID: 10515153]

[80] Meuer SC, Dumann H, Meyer zum Büschenfelde KH, Köhler H. Low-dose interleukin-2 induces systemic immune responses against HBsAg in immunodeficient non-responders to hepatitis B vaccination. Lancet 1989; 1(8628): 15-8.
[http://dx.doi.org/10.1016/S0140-6736(89)91674-7] [PMID: 2563003]

[81] Zwahlen RD, Spreng D, Wyder-Walther M. *In vitro* and *in vivo* activity of human interleukin-8 in dogs. Vet Pathol 1994; 31(1): 61-6.
[http://dx.doi.org/10.1177/030098589403100108] [PMID: 8140727]

[82] Sadler AJ, Williams BR. Interferon-inducible antiviral effectors. Nat Rev Immunol 2008; 8(7): 559-68.
[http://dx.doi.org/10.1038/nri2314] [PMID: 18575461]

[83] Sirén J, Pirhonen J, Julkunen I, Matikainen S. IFN-alpha regulates TLR-dependent gene expression of IFN-alpha, IFN-beta, IL-28, and IL-29. J Immunol 2005; 174(4): 1932-7.
[http://dx.doi.org/10.4049/jimmunol.174.4.1932] [PMID: 15699120]

[84] Colonna M, Pulendran B, Iwasaki A. Dendritic cells at the host-pathogen interface. Nat Immunol 2006; 7(2): 117-20.
[http://dx.doi.org/10.1038/ni0206-117] [PMID: 16424884]

[85] Bracklein T, Theise S, Metzler A, Spiess BM, Richter M. Activity of feline interferon-omega after ocular or oral administration in cats as indicated by Mx protein expression in conjunctival and white blood cells. Am J Vet Res 2006; 67(6): 1025-32.
[http://dx.doi.org/10.2460/ajvr.67.6.1025] [PMID: 16740097]

[86] Goodbourn S, Didcock L, Randall RE. Interferons: cell signalling, immune modulation, antiviral response and virus countermeasures. J Gen Virol 2000; 81(Pt 10): 2341-64.
[PMID: 10993923]

[87] Lee S, Margolin K. Cytokines in cancer immunotherapy. Cancers (Basel) 2011; 3(4): 3856-93.
[http://dx.doi.org/10.3390/cancers3043856] [PMID: 24213115]

[88] Fulton RW, Burge LJ. Susceptibility of feline herpesvirus 1 and a feline calicivirus to feline interferon and recombinant human leukocyte interferons. Antimicrob Agents Chemother 1985; 28(5): 698-9.
[http://dx.doi.org/10.1128/AAC.28.5.698] [PMID: 2418778]

[89] Jameson P, Essex M. Inhibition of feline leukemia virus replication by human leukocyte interferon. Antiviral Res 1983; 3(2): 115-20.
[http://dx.doi.org/10.1016/0166-3542(83)90033-5] [PMID: 6312876]

[90] Pedretti E, Passeri B, Amadori M, *et al.* Low-dose interferon-alpha treatment for feline immunodeficiency virus infection. Vet Immunol Immunopathol 2006; 109(3-4): 245-54.
[http://dx.doi.org/10.1016/j.vetimm.2005.08.020] [PMID: 16169599]

[91] Weiss RC, Oostrom-Ram T. Inhibitory effects of ribavirin alone or combined with human alpha interferon on feline infectious peritonitis virus replication *in vitro.* Vet Microbiol 1989; 20(3): 255-65.
[http://dx.doi.org/10.1016/0378-1135(89)90049-7] [PMID: 2549687]

[92] Zeidner NS, Myles MH, Mathiason-DuBard CK, Dreitz MJ, Mullins JI, Hoover EA. Alpha interferon (2b) in combination with zidovudine for the treatment of presymptomatic feline leukemia virus-induced immunodeficiency syndrome. Antimicrob Agents Chemother 1990; 34(9): 1749-56.
[http://dx.doi.org/10.1128/AAC.34.9.1749] [PMID: 2178336]

[93] Schellekens H, Geelen G, Meritet JF, Maury C, Tovey MG. Oromucosal interferon therapy: relationship between antiviral activity and viral load. J Interferon Cytokine Res 2001; 21(8): 575-81.
[http://dx.doi.org/10.1089/10799900152547830] [PMID: 11559435]

[94] Tompkins WA. Immunomodulation and therapeutic effects of the oral use of interferon-alpha: mechanism of action. J Interferon Cytokine Res 1999; 19(8): 817-28. [http://dx.doi.org/10.1089/107999099313325] [PMID: 10476925]

[95] Collado VM, Gómez-Lucía E, Tejerizo G, *et al.* Effect of type I interferons on the expression of feline leukaemia virus. Vet Microbiol 2007; 123(1-3): 180-6. [http://dx.doi.org/10.1016/j.vetmic.2007.03.008] [PMID: 17507184]

[96] Cummins JM, Tompkins MB, Olsen RG, Tompkins WA, Lewis MG. Oral use of human alpha interferon in cats. J Biol Response Mod 1988; 7(5): 513-23. [PMID: 3183686]

[97] Ueda Y, Sakurai T, Kasama K, *et al.* Pharmacokinetic properties of recombinant feline interferon and its stimulatory effect on 2′,5′-oligoadenylate synthetase activity in the cat. J Vet Med Sci 1993; 55(1): 1-6. [http://dx.doi.org/10.1292/jvms.55.1] [PMID: 7681698]

[98] Mochizuki M, Nakatani H, Yoshida M. Inhibitory effects of recombinant feline interferon on the replication of feline enteropathogenic viruses *in vitro*. Vet Microb 1994; 39(1-2): 145-52.

[99] Siebeck N, Hurley DJ, Garcia M, *et al.* Effects of human recombinant alpha-2b interferon and feline recombinant omega interferon on *in vitro* replication of feline herpesvirus-1. Am J Vet Res 2006; 67(8): 1406-11. [http://dx.doi.org/10.2460/ajvr.67.8.1406] [PMID: 16881854]

[100] Tateyama S, Priosoeryanto BP, Yamaguchi R, Uchida K, Ogiwara K, Suchiya AT. *In vitro* growth inhibition activities of recombinant feline interferon on all lines derived from canine tumours. Res Vet Sci 1995; 59(3): 275-7. [http://dx.doi.org/10.1016/0034-5288(95)90017-9] [PMID: 8588106]

[101] Haid C, Kaps S, Gönczi E, *et al.* Pretreatment with feline interferon omega and the course of subsequent infection with feline herpesvirus in cats. Vet Ophthalmol 2007; 10(5): 278-84. [http://dx.doi.org/10.1111/j.1463-5224.2007.00550.x] [PMID: 17760705]

[102] Hennet PR, Camy GA, McGahie DM, Albouy MV. Comparative efficacy of a recombinant feline interferon omega in refractory cases of calicivirus-positive cats with caudal stomatitis: a randomised, multi-centre, controlled, double-blind study in 39 cats. J Feline Med Surg 2011; 13(8): 577-87. [http://dx.doi.org/10.1016/j.jfms.2011.05.012] [PMID: 21802033]

[103] Ritz S, Egberink H, Hartmann K. Effect of feline interferon-omega on the survival time and quality of life of cats with feline infectious peritonitis. J Vet Intern Med 2007; 21(6): 1193-7. [http://dx.doi.org/10.1111/j.1939-1676.2007.tb01937.x] [PMID: 18196725]

[104] Priosoeryanto BP, Tateyama S, Yamaguchi R, Uchida K. Antiproliferation and colony-forming inhibition activities of recombinant feline interferon (rFeIFN) on various cells *in vitro*. Can J Vet Res 1995; 59(1): 67-9. [PMID: 7704845]

[105] Litzlbauer P, Weber K, Mueller RS. Oral and subcutaneous therapy of canine atopic dermatitis with recombinant feline interferon omega. Cytokine 2014; 66(1): 54-9. [http://dx.doi.org/10.1016/j.cyto.2013.12.001] [PMID: 24548425]

[106] Leal RO, Gil S, Brito MT, McGahie D, Niza MM, Tavares L. The use of oral recombinant feline interferon omega in two cats with type II diabetes mellitus and concurrent feline chronic gingivostomatitis syndrome. Ir Vet J 2013; 66(1): 19.
[http://dx.doi.org/10.1186/2046-0481-66-19] [PMID: 24153100]

[107] Ishiwata K, Minagawa T, Kajimoto T. Clinical effects of the recombinant feline interferon-omega on experimental parvovirus infection in beagle dogs. J Vet Med Sci/ Jpn J Vet Sci 1998; 60(8): 911-7.
[http://dx.doi.org/10.1292/jvms.60.911]

[108] de Mari K, Maynard L, Sanquer A, Lebreux B, Eun HM. Therapeutic effects of recombinant feline interferon-omega on feline leukemia virus (FeLV)-infected and FeLV/feline immunodeficiency virus (FIV)-coinfected symptomatic cats. J Vet Intern Med 2004; 18(4): 477-82.
[http://dx.doi.org/10.1892/0891-6640(2004)18<477:TEORFI&;2.0.CO;2] [PMID: 15320583]

[109] Doménech A, Miró G, Collado VM, *et al.* Use of recombinant interferon omega in feline retrovirosis: from theory to practice. Vet Immunol Immunopathol 2011; 143(3-4): 301-6.
[http://dx.doi.org/10.1016/j.vetimm.2011.06.008] [PMID: 21719116]

[110] Gil S, Leal RO, Duarte A, *et al.* Relevance of feline interferon omega for clinical improvement and reduction of concurrent viral excretion in retrovirus infected cats from a rescue shelter. Res Vet Sci 2013; 94(3): 753-63.
[http://dx.doi.org/10.1016/j.rvsc.2012.09.025] [PMID: 23122808]

[111] Leal RO, Gil S, Sepúlveda N, *et al.* Monitoring acute phase proteins in retrovirus infected cats undergoing feline interferon-ω therapy. J Small Anim Pract 2014; 55(1): 39-45.
[http://dx.doi.org/10.1111/jsap.12160] [PMID: 24279640]

[112] Gil S, Leal RO, McGahie D, *et al.* Oral Recombinant Feline Interferon-Omega as an alternative immune modulation therapy in FIV positive cats: clinical and laboratory evaluation. Res Vet Sci 2014; 96(1): 79-85.
[http://dx.doi.org/10.1016/j.rvsc.2013.11.007] [PMID: 24332273]

[113] Bowles CA, Messerschmidt G, Deisseroth AB. Experience with *Staphylococcus aureus* and protein A in the treatment of canine malignancies. J Biol Response Mod 1984; 3(3): 260-5.
[PMID: 6379110]

[114] Klausner JS, Miller WJ, O'Brien TD, Branda RF. Effects of plasma treatment with purified protein A and Staphylococcus aureus Cowan I on spontaneous animal neoplasms. Cancer Res 1985; 45(3): 1263-9.
[PMID: 3971372]

[115] Harper HD, Sjöquist J, Hardy WD Jr, Jones FR. Antitumor activity of protein A administered intravenously to pet cats with leukemia or lymphosarcoma. Cancer 1985; 55(9): 1863-7.
[http://dx.doi.org/10.1002/1097-0142(19850501)55:9<1863::AID-CNCR2820550903>3.0.CO;2-T] [PMID: 2983866]

[116] Lafrado LJ, Mathes LE, Zack PM, Olsen RG. Biological effects of staphylococcal protein A immunotherapy in cats with induced feline leukemia virus infection. Am J Vet Res 1990; 51(3): 482-6.
[PMID: 1690524]

[117] McCaw DL, Boon GD, Jergens AE, Kern MR, Bowles MH, Johnson JC. Immunomodulation therapy

for feline leukemia virus infection. J Am Anim Hosp Assoc 2001; 37(4): 356-63.
[http://dx.doi.org/10.5326/15473317-37-4-356] [PMID: 11450836]

[118] Megid J, Kaneno R. Natural killer activity in mice infected with rabies virus and submitted to P. acnes (Propionibacterium acnes) as immunomodulator. Comp Immunol Microbiol Infect Dis 2000; 23(2): 91-7.
[http://dx.doi.org/10.1016/S0147-9571(99)00061-2] [PMID: 10670698]

[119] Theilen GH, Hills D. Comparative aspects of cancer immunotherapy: immunologic methods used for treatment of spontaneous cancer in animals. J Am Vet Med Assoc 1982; 181(10): 1134-41.
[PMID: 6757216]

[120] Becker AM, Janik TA, Smith EK, Sousa CA, Peters BA. Propionibacterium acnes immunotherapy in chronic recurrent canine pyoderma. An adjunct to antibiotic therapy. J Vet Intern Med 1989; 3(1): 26-30.
[http://dx.doi.org/10.1111/j.1939-1676.1989.tb00325.x] [PMID: 2647969]

[121] Weiss RC, Cox NR, Oostrom-Ram T. Effect of interferon or Propionibacterium acnes on the course of experimentally induced feline infectious peritonitis in specific-pathogen-free and random-source cats. Am J Vet Res 1990; 51(5): 726-33.
[PMID: 2159739]

[122] Harris C, Pierce K, King G, Yates KM, Hall J, Tizard I. Efficacy of acemannan in treatment of canine and feline spontaneous neoplasms. Mol Biother 1991; 3(4): 207-13.
[PMID: 1768373]

[123] King GK, Yates KM, Greenlee PG, *et al.* The effect of Acemannan Immunostimulant in combination with surgery and radiation therapy on spontaneous canine and feline fibrosarcomas. J Am Anim Hosp Assoc 1995; 31(5): 439-47.
[http://dx.doi.org/10.5326/15473317-31-5-439] [PMID: 8542364]

[124] Yates KM, Rosenberg LJ, Harris CK, *et al.* Pilot study of the effect of acemannan in cats infected with feline immunodeficiency virus. Vet Immunol Immunopathol 1992; 35(1-2): 177-89.
[http://dx.doi.org/10.1016/0165-2427(92)90130-I] [PMID: 1337396]

[125] Sheets MA, Unger BA, Giggleman GF Jr, Tizard IR. Studies of the effect of acemannan on retrovirus infections: clinical stabilization of feline leukemia virus-infected cats. Mol Biother 1991; 3(1): 41-5.
[PMID: 1648929]

[126] MacEwen EG. An immunologic approach to the treatment of cancer. Vet Clin North Am 1977; 7(1): 65-75.
[http://dx.doi.org/10.1016/S0091-0279(77)50006-8] [PMID: 191975]

[127] Bachman D, Cockerell GL, de Noronha F, Rapp HJ. Effect of bacille Calmette-Guerin immunotherapy on feline sarcoma virus-induced neoplasms in the cat. Am J Vet Res 1982; 43(3): 475-80.
[PMID: 6280529]

[128] Armstrong PJ. Systemic Serratia marcescens infections in a dog and a cat. J Am Vet Med Assoc 1984; 184(9): 1154-8.
[PMID: 6373698]

[129] Elmslie RE, Ogilvie GK, Dow SW, *et al.* Evaluation of a biologic response modifier derived from

Serratia marcescens: effects on feline macrophages and usefulness for the prevention and treatment of viremia in feline leukemia virus-infected cats. Mol Biother 1991; 3(4): 231-8.
[PMID: 1768375]

[130] Ogilvie GK, Elmslie RE, Cecchini M, Walters LM, Pearson FC. Use of a biological extract of Serratia marcescens to decrease doxorubicin-induced myelosuppression in dogs. Am J Vet Res 1992; 53(10): 1787-90.
[PMID: 1280926]

[131] Hörber D, Mayr B. [Paramunization of FeLV-positive cats with PIND-AVI]. Tierarztl Prax 1991; 19(3): 311-4.
[PMID: 1653471]

[132] Hartmann K, Block A, Ferk G, Vollmar A, Goldberg M, Lutz H. Treatment of feline leukemia virus-infected cats with paramunity inducer. Vet Immunol Immunopathol 1998; 65(2-4): 267-75.
[http://dx.doi.org/10.1016/S0165-2427(98)00161-5] [PMID: 9839879]

[133] Hartmann K, Block A, Ferk G, Beer B, Vollmar A, Lutz H. Treatment of feline leukemia virus (FeLV) infection. Vet Microbiol 1999; 69(1-2): 111-3.
[http://dx.doi.org/10.1016/S0378-1135(99)00097-8] [PMID: 10515279]

[134] Reinemeyer CR, Faulkner CT, Assadi-Rad AM, Burr JH, Patton S. Comparison of the efficacies of three heartworm preventives against experimentally induced infections with Ancylostoma caninum and Toxocara canis in pups. J Am Vet Med Assoc 1995; 206(11): 1710-5.
[PMID: 7782242]

[135] MacEwen EG, Hayes AA, Mooney S, *et al.* Evaluation of effect of levamisole on feline mammary cancer. J Biol Response Mod 1984; 3(5): 541-6.
[PMID: 6502187]

[136] Hsu WH. Toxicity and drug interactions of levamisole. J Am Vet Med Assoc 1980, 176(10 Spec No) 1980; 1166-9.

[137] Yamauchi K, Wakabayashi H, Hashimoto S, Teraguchi S, Hayasawa H, Tomita M. Effects of orally administered bovine lactoferrin on the immune system of healthy volunteers. Adv Exp Med Biol 1998; 443: 261-5.
[http://dx.doi.org/10.1007/978-1-4757-9068-9_32] [PMID: 9781368]

[138] Sato R, Inanami O, Tanaka Y, Takase M, Naito Y. Oral administration of bovine lactoferrin for treatment of intractable stomatitis in feline immunodeficiency virus (FIV)-positive and FIV-negative cats. Am J Vet Res 1996; 57(10): 1443-6.
[PMID: 8896681]

[139] WH (2009). Safety issues in the preparation of homeopathic medicines. Obtido de http://apps.who.int//iris/handle/10665/44238 2009.

[140] Wynn SG. Studies on use of homeopathy in animals. J Am Vet Med Assoc 1998; 212(5): 719-24.
[PMID: 9524649]

[141] Mellman I, Coukos G, Dranoff G. Cancer immunotherapy comes of age. Nature 2011; 480(7378): 480-9.
[http://dx.doi.org/10.1038/nature10673] [PMID: 22193102]

[142] Gulley JL, Madan RA, Schlom J. Impact of tumour volume on the potential efficacy of therapeutic vaccines. Curr Oncol 2011; 18(3): e150-7.
[http://dx.doi.org/10.3747/co.v18i3.783] [PMID: 21655153]

[143] Schlom J. Therapeutic cancer vaccines: current status and moving forward. J Natl Cancer Inst 2012; 104(8): 599-613.
[http://dx.doi.org/10.1093/jnci/djs033] [PMID: 22395641]

[144] Emens LA. Cancer vaccines: on the threshold of success. Expert Opin Emerg Drugs 2008; 13(2): 295-308.
[http://dx.doi.org/10.1517/14728214.13.2.295] [PMID: 18537522]

[145] Schlom J, Hodge J, Palena C, *et al.* Therapeutic Cancer Vaccines. Tew K, Fisher P, Advances in Cancer Research. United States of America: Elsevier 2014; Vol. 121.

[146] Berzofsky JA, Terabe M, Oh S, *et al.* Progress on new vaccine strategies for the immunotherapy and prevention of cancer. J Clin Invest 2004; 113(11): 1515-25.
[http://dx.doi.org/10.1172/JCI21926] [PMID: 15173875]

[147] Harrop R, John J, Carroll MW. Recombinant viral vectors: cancer vaccines. Adv Drug Deliv Rev 2006; 58(8): 931-47.
[http://dx.doi.org/10.1016/j.addr.2006.05.005] [PMID: 17030074]

[148] Fong L, Engleman EG. Dendritic cells in cancer immunotherapy. Annu Rev Immunol 2000; 18: 245-73.
[http://dx.doi.org/10.1146/annurev.immunol.18.1.245] [PMID: 10837059]

[149] Gilboa E. DC-based cancer vaccines. J Clin Invest 2007; 117(5): 1195-203.
[http://dx.doi.org/10.1172/JCI31205] [PMID: 17476349]

[150] El-Amm J, Aragon-Ching JB. The changing landscape in the treatment of metastatic castration-resistant prostate cancer. Ther Adv Med Oncol 2013; 5(1): 25-40.
[http://dx.doi.org/10.1177/1758834012458137] [PMID: 23323145]

[151] Schorey JS, Bhatnagar S. Exosome function: from tumor immunology to pathogen biology. Traffic 2008; 9(6): 871-81.
[http://dx.doi.org/10.1111/j.1600-0854.2008.00734.x] [PMID: 18331451]

[152] Pardoll DM. The blockade of immune checkpoints in cancer immunotherapy. Nat Rev Cancer 2012; 12(4): 252-64.
[http://dx.doi.org/10.1038/nrc3239] [PMID: 22437870]

[153] Guo C, Manjili MH, Subjeck JR, Sarkar D, Fisher PB, Wang XY. Therapeutic cancer vaccines: past, present, and future. Adv Cancer Res 2013; 119: 421-75.
[http://dx.doi.org/10.1016/B978-0-12-407190-2.00007-1] [PMID: 23870514]

CHAPTER 6

Perspectives on Natural Products

Ana Patrícia Quendera, Teresa Semedo-Lemsaddek *

Centro de Investigação Interdisciplinar em Sanidade Animal, Faculdade de Medicina Veterinária, Universidade de Lisboa, Avenida da Universidade Técnica, 1300-477 Lisboa, Portugal

Abstract: The emergence of antibiotic resistant strains along with the increasing public health concern regarding infectious diseases and hospital-acquired infections has led to the necessity of finding viable alternatives to antibiotics.

Nature has been a source of compounds with interesting medicinal properties for millennia. For many decades, natural products have been a wealthy source of antimicrobials and more recently academic drug discovery in this research area has been accentuated, though pharmaceutical industries devote fewer resources to antimicrobial drug discovery programs, in comparison with their investment several decades ago.

Though, initially, the developments in drug discovery were aimed towards synthetic chemical libraries, nowadays they are also greatly used in the research of bioactive natural products, in order to keep up with the developments in similar areas. Moreover, these developments allowed to overcome several problems associated with natural products drug discovery by using molecular techniques (i.e. genome mining) when organisms are not cultivable on the laboratorial environment.

In this chapter, various examples of natural products with antimicrobial properties from marine organisms and plants are referred, highlighting their important and promising results against pathogenic bacteria, fungi, viruses and protozoa.

Keywords: Antibacterial, Antibiotics, Antifungal, Antiprotozoal, Antiviral, Bacteria, Bioactive molecules, Drug discovery, Fungi, Infectious diseases, Marine organisms, Natural products, Pathogenic microorganisms, Plants, Secondary metabolite.

+ **Corresponding author Teresa Semedo-Lemsaddek:** Centro de Investigação Interdisciplinar em Sanidade Animal (CIISA), Faculdade de Medicina Veterinária da Universidade de Lisboa, Lisboa, Portugal; E-mail: tlemsaddek@fmv.ulisboa.pt

Manuela Oliveira and Isa Serrano (Eds)

1. NATURAL PRODUCTS AS ANTIBIOTIC ALTERNATIVES FOR THE NEW MILLENNIUM

Natural products are frequently referred to as secondary metabolites produced by living organisms and are composed of small molecules which lack an apparent function in growth, development, reproduction or survival processes on the producing organism. Instead, these secondary metabolites have attracted attention because of their biological effect on other organisms [1].

Initially regarded as waste products, natural compounds often present an ecological role in the regulation of the interactions between animals, plants and microorganisms, acting as defensive substances against predation and consumption, attractants or pheromones [2, 3].

Hence, natural products can be (i) entire organisms, either plant, animal or microorganisms, which have not been subjected to any processing besides its preservation; (ii) part of an organism, such as leaves of a plant or an animal organ; (iii) an extract of an organism or an exudate or (iv) pure compounds isolated from plants, animals or microorganisms [1].

Since ancient human history, natural products have been used for healing and relieving diseases. For example, the use of bearberry (*Arctostaphylos uva-ursi*) and cranberry juice (*Vaccinium macrocarpon*) to treat urinary tract infections is described in different manuals of phytotherapy; on the other hand, lemon balm (*Melissa officinalis*), garlic (*Allium sativum*) and tea tree (*Melaleuca alternifolia*) are reported as broad spectrum antimicrobial agents, in particular tea tree essential oil is used to treat acne and other skin infections [4].

Based on their use in traditional medicine, a large number of modern drugs have been derived from natural sources. A quarter to half of currently available drugs was originated from natural compounds [5]. Naturally occurring molecules serve as the source for the majority of the medicaments used in clinical settings and more than 50% of Food and Drug Administration (FDA)-approved drugs are either natural products or derivatives. Relatively to disease categories, approximately 68% of anti-infective (anti-bacterial, -fungal, -parasitic and -viral) were classified as naturally-derived or naturally-inspired, whereas in cancer treatment these numbers rise above 79% [6].

Their structural diversity along with their unique and complex structures has long been appreciated, being very important scaffolds in drug discovery research. Moreover, the fact that most natural resources remain unknown and undiscovered is remarkable. Only 6% of the 300,000 species of higher terrestrial plants have been investigated in the pharmacological and chemical area. The same happens in the marine environment, where most areas remain unexplored [2].

Since the late 1980s, the average number of new chemical entities has been decreasing from over 60 small molecules per year to only 23, from 2001 to 2010 [6]. The lack of interest in natural products by major pharmaceutical companies in favor of new chemical techniques was one of the factors responsible for this declining in the research on natural products [7].

Nevertheless, for years to come, natural products will still be relevant in new drug research, due to their inherent unparalleled chemical structural diversity and qualifications regarding "(i) the rate of introduction of new chemical entities of wide structural diversity, including serving as templates for semisynthetic and total synthetic modification; (ii) the number of diseases which can be treated or prevented by these substances; and (iii) their frequency of use" [8]. Additionally, these compounds present a tendency to interact with biological targets, meaning they are suitable lead structures.

2. SEARCHING FOR NATURAL PRODUCTS: THE IMPORTANCE AND EVOLVING ROLE OF NATURAL PRODUCTS IN DRUG DISCOVERY

Natural compounds have made and continue to make a unique contribution for the discovery and development of effective drugs to be used for the treatment of various human diseases. They can be applied as new drugs to be readily used in an unmodified state (*e.g.* vincristine from *Catharanthus roseus*), as chemical scaffolds used to synthesize other complex molecules (*e.g.* diosgenin from *Dioscorea floribunda* for the synthesis of oral contraceptives) or as indicators of new ways of pharmacological action in order to synthesize new analogs (*e.g.* synthetic analogs of penicillin from *Penicillium notatum*) [1].

Multidisciplinary collaboration plays an essential role in drug discovery and product development. Biologists provide suitable disease-oriented screening studies, which lead to the discovery of bioactive molecules. In many cases, such compound is not suitable for development into an effective drug, being regarded as a lead molecule which can build the basis for further chemical or biochemical modification, involving biochemists, medicinal and synthetic chemists. Ultimately, in order to determine the optimal pharmacodynamic and toxicological parameters needed for further proceeding into clinical trials, collaboration with pharmacologist and toxicologists is required [9].

Natural compounds question a different area of chemical space than synthetic compounds, since there are significant differences in the molecular structure produced by nature in comparison with synthetic molecules. Natural structures take their limited building blocks and break them into panoply of pathways while synthetic chemistry uses repetitions of a sequence of reactions and only changes the input [10].

Moreover, natural products are responsive to further improvement, either in efficacy and target selectivity or by achieving optimal pharmacokinetic and pharmacodynamic properties, since therapeutic use is not their evolutionary basis of existence. For example, the opium alkaloid morphine is a natural product used in extract and pure form and has led to the discovery of many semisynthetic and fully synthetic compounds which are considered good second-generation opioid drugs [10].

For the last decade, the reduced interest on the discovery of natural products by the pharmaceutical industry could be linked to several reasons. In first place, high-throughput screening systems against defined molecular targets were introduced, shifting the interest of many companies to synthetic chemical libraries. Secondly, there was the development of combinatorial chemistry, providing simpler and more drug-like screening libraries. Third, advances in molecular biology, cellular biology and genomics multiplied the number of molecular targets and spurred shorter drug discovery timelines. Lastly, pharmaceutical companies started neglecting research on infectious disease therapies, which was the ultimate strength of natural compounds [11, 12].

Hence, these factors clearly show that present-day drug research wants rapid screening, hit identification and hit-to-lead development [13]. As stated by Gray *et al.* (2012) [11], "in the industrialist point of view, intensive traditional resource natural-product programs based on extract-library screening, bioassay-guided isolation, structure elucidation and production scale-up face a distinct competitive disadvantage when compared with approaches that utilize defined synthetic chemical libraries". Nonetheless, emerging trends and unexpected outcomes from current research are suggesting a novel interest in the unexploited microbial, plant and animal natural compounds as "a source of chemical diversity and lead generation" [13 - 16]. The urgent necessity of new pharmaceuticals for the treatment of cancer and infectious diseases was also a major encouragement to pursuit the investigation on natural products.

Besides structural diversity and complexity, natural products often possess highly selective and specific biological activities based on mechanisms of action. Their bioactivity is on the basis of the assumption that all natural compounds have some receptor-binding activity. Often, organisms possess complex arrays of unique bioactive constituents, analogous to the libraries of crude synthetic products produced by combinatorial chemistry techniques, which could be used to optimize the structures of natural compounds in order to achieve compounds of pharmacological interest [7].

3. CONSIDERATIONS FOR THE IDENTIFICATION, ISOLATION AND TESTING OF NATURAL PRODUCTS AND THEIR ACTIVE COMPOUNDS

For the last decades, research strategies in natural products area have grown considerably in order to follow the same pace as the developments in other similar areas of drug discovery. Traditional methods started with the extraction of the natural product from the source, followed by screening against a specific target, isolation, fractionation and purification, in order to yield a single biological active compound. Though they are still used and remain valid, traditional methods are slow, demand more labor and success lacks guarantees [1, 17].

The modern approaches are based on the following premises: bioassay-guided isolation and identification of active main components; production of dereplicated natural products libraries for high-throughput system screening; production of active compounds in cell or tissue culture, genetic manipulation and natural combinatorial chemistry; major focus on bioactivity; introduction of the concepts of chemical fingerprinting and metabolomics; selection of organisms with basis on ethnopharmacological information and randomly selected; broadening the range of natural sources (*e.g.* marine organisms, insects) [1].

Currently, various methods for the extraction and isolation of natural products are available. The design of the protocol can only be performed when the target compound or the general purpose has been established, and information about the chemical and physical nature of the product in question is relevant. When the compounds are unknown it may be necessary to, previously, test out extraction and isolation methods in order to select the best option, having in mind the advantages and disadvantages, their efficiency and involved costs. These initial steps may be followed by structure determination or confirmation of the identity of the purified components. Due to the progresses in the area of mass spectrometry and nuclear magnetic resonance, nowadays the structure of a compound is possible to deduce based on microgram amounts [1].

High-throughput screening systems have become crucial in modern drug discovery from natural products. It corresponds to a standardized and replicate screening method which utilizes 96-well microplates to screen samples. Using a multilabel counter and suitable indicators, biological activity is determined through fluorescence, luminescence, optical density or by radiometric detectors in order to give accurate and reproducible results [11].

The primary purpose of high-throughput screening systems is to identify a diverse set of chemical hits, which are divided into broad chemical series evaluated according to "their potency, physical and chemical properties, predicted ease of synthesis, scalability, target selectivity, structural alerts, freedom to operate and the ability to create and protect intellectual property" [11]. From finding a hit until the main compound is identified, it takes an evaluation of those chemical series which survived the triage to identify potential liabilities linked with each series.

The development of analytical and spectroscopic methods, such as nuclear magnetic resonance and mass spectrometry, has contributed to bring natural compounds back to the path of drug research [13]. In particular, advances in spectroscopic methods led to the *de novo* structure determination of new chemical entities in rather small concentrations and in complex mixtures. Additionally, recent advances allowed the access to microbial sources unattainable until then. Being independent from culture media, methods based only on the total DNA of the bacterial community, namely metagenomics, enabled the discovery of uncultivable organisms [18, 19].

Genome mining is another novel technique which consists in the analysis of genome sequences for the identification of genes encoding proteins [17]. This technique is an additional approach to discover new natural compounds and it has revealed gene clusters and novel pathways for the biosynthesis of several known and unknown natural compounds. The genes encoding natural products (including antibiotics) produced by bacteria are usually closely clustered along the genome, which facilitates the identification of the number and type of natural products produced by an organism. *B. subtilis*, for example, is an antibiotic producer which has been studied for more than a century, and the complete genome sequence of strain 168 revealed a new 80 kb cluster for natural products biosynthesis, initially considered to be non-functional [20]. Recently, researchers identified the structure of the polyketide-peptide hybrid compound bacillaene, a bacteriostatic agent produced by *B. subtilis* which inhibits prokaryotic protein synthesis [21]. By knowing the structure of bacillaene, the study of its biosynthetic steps is facilitated and it is now possible to explore the biological role of these compounds and even to produce structural analogs [22]. In the same way, coelichelin was shown to be produced by the model actinomycete *Streptomyces coelicolor*, a new siderophore responsible for its iron uptake [23, 24]. Its genome sequence also revealed potential genes clusters for *circa* 20 natural products [25].

Genome mining also permits the expression of gene clusters in heterologous organism in order to obtain the secondary metabolites desired, and combinatorial biosynthesis, the genetic manipulation of the gene cluster involved in natural products synthesis to get new molecules [26]. In addition, with the ultimate advances in bioinformatics, it is now possible to predict the chemical structure of the natural product produced by the biosynthetic gene cluster. One example is the novel antimicrobial and surfactant cyclic lipopeptide from *Pseudomonas fluorescens* SBW25 (a plant beneficial strain), having a role in motility, biofilm formation and lytic activity against zoospores of *Phytophthora infestans*, a plant pathogen [27].

Research in novel antimicrobial agents from plants has been facing several inherent difficulties. The lack of standardized *in vitro* methods for testing antimicrobial activities of plant crude extracts leads to variations in the results between different research groups which compromises direct comparisons and also leads to false conclusions on the antimicrobial efficacy of the compounds under study [28].

The Clinical and Laboratory Standards Institute (CLSI) guidelines for the evaluation of antimicrobial compounds are widely accepted by regulatory bodies worldwide [29]. Nonetheless, these methods were elaborated to test pure antimicrobial molecules, such as antibiotics, adjustments being necessary before the use of plant extracts for antimicrobial assays [30].

In addition, comparison of obtained results with published ones is usually difficult since many factors influence the antimicrobial activity of the plant extract in study: environment and climatic conditions during plant growth, choice of plant extracts, extraction methodology, antimicrobial assays, and tested microorganisms [29, 30].

The most used methods for the assessment of antimicrobial activities include disc diffusion, well diffusion and broth dilution, either individually or together, being the first method the most often employed.

Since each bacterial strain has a different growth rate, the growth curve of each strain should be established to target the logarithmic phase and to assess the initial number of bacteria (inoculums) to be used in the following assays [28]. The standard inoculum size often is of 10^8 Colonies Forming Units (CFU)/ml of bacteria, which is equivalent to McFarland 0.5 turbidity standard [29].

The agar disc diffusion assay is a modification of the Kirby-Bauer test, using 6 mm sterilized filter paper discs which are then soaked in filter sterilized plant extracts. The impregnated discs are placed onto the surface of an agar medium plate, previously inoculated with the test microorganism. Several types of solid agar medium are used, such as Mueller-Hinton, tryptone soy agar, nutrient agar or even brain-heart infusion agar [29]. Other researchers impregnate the paper discs after placing onto the plates. The drying time of the discs also varies among assays; temperature and time of incubation could be the optimal growth temperature for used microorganisms, or 24 h at 37°C for bacteria and 48 h at 25°C for fungi. After incubation, diameter of the zone of inhibition, *i.e.*, without visible growth, is measured to the nearest millimeter.

Regarding the well diffusion assay, agar plates are inoculated and spread with a standardized concentration of inoculum, like the disc diffusion assay. Then, holes with 6-8 mm wide are punched with the upper side of an aseptic Pasteur pipette, and those holes are filled with a fixed volume of plant extract at the desired concentration. Temperature and time of incubation are equivalent to agar disc diffusion.

Diffusion assays are simple and rapid methods which require rather small amounts of plant extracts. They are regularly used as preliminary tests for the antimicrobial activity of a large number of compounds, in order to select the ones with the highest inhibitory activity, *i.e.*, larger inhibitory zones, especially disc diffusion method [31].

However, diffusion methods present some disadvantages. First, using volatile compounds could lead to reduced zones of inhibition, since they could evaporate very quickly. Also, poorly soluble compounds do not diffuse uniformly through the agar medium plates [32]. Secondly, for non-polar extracts and essential oils, diffusion techniques are not suitable for testing [4, 33]. Therefore, the convenience of disc diffusion method is limited to the attainment of preliminary and qualitative data, turning essential the performance of other complementary methodologies.

Broth microdilution method can be used to further evaluate the antimicrobial activity of the most efficient plant extracts and is commonly used to determine the extent of inhibitory activity, namely by determining the minimum inhibitory concentration (MIC) and the minimum bactericidal concentration (MBC). MIC, which corresponds to the lowest concentration of antimicrobials which inhibit the visible growth after incubation, is also a quantitative endpoint measurement most commonly used for evaluating the antimicrobial effect of antibiotics. Nonetheless, MIC and MBC definitions may slightly vary between different researchers, turning even more problematic the comparison of results.

Determination of MICs for essential oils and extracts which are insoluble in water and broth media has been problematic. Broth microdilution assays have relied mostly on emulsifiers or solvents, like Tween 20 (polyoxyethylene (2) sorbitan mono-laurate) and Tween 80 (polysorbate 80), dimethysulfoxide (DMSO) or even ethanol, to guarantee contact between the test organism and the antimicrobial agent during the entire experiment. However, these agents may interfere with the physical and chemical properties of the system, and either increasing or diminishing the antimicrobial effect. Mann and Markham [33] reported the use of 0.15% agar as the best solvent to overcome the problem of emulsion stability for tea tree oil, instead of the four agents previously mentioned.

4. PROMISING RESULTS ON THE USAGE OF NATURAL COMPOUNDS AGAINST PATHOGENIC MICROORGANISMS

In the last decade, research towards the discovery of antimicrobial agents derived from all sorts of natural products has been accentuated, though pharmaceutical industries devote fewer resources to antimicrobial drug discovery programs in comparison with their investment during the 'golden age' [16]. The major reasons being the emergence of infectious diseases [34] and the increasing public health concern of hospital-acquired infections caused by bacteria which are resistant to more than three classes of conventionally used antibiotics. The discovery of new antibiotics and advances in antimicrobial drug research area led to the improvement of human health and more efficient treatments of infections [35].

Nonetheless, the worldwide excessive and extended use and misuse of antibiotics in human/veterinary medicine, animal production and agriculture explains the growing number of resistant bacteria, probably due to elimination of susceptible strains and selection of resistant variants, in particular multidrug resistant ones [36, 37]. This situation has significantly decreased the efficiency of antibiotics, demanding for alternative strategies to combat microbial infections. Moreover, the antibiotics currently in use were created based on limited chemical scaffolds, with minor innovations since the 1980s, which facilitates the increasing development of microbial wide-range resistance mechanisms [38, 39].

One of the utmost problems in antibiotic-drug-discovery is the difficulty in identifying new drug-like compounds with suitable antimicrobial activity. The large majority of the antibiotics used in clinical settings were found as secondary metabolites produced from microorganisms, mostly fungi and members of the Actinobacteria phylum, in particular within the genus *Streptomyces*. This genus is responsible for the production of *circa* 100,000 secondary metabolites with some antimicrobial activity [40]. Yet, the frequency of identification of novel antimicrobial compounds has decreased since the 1970s because the undiscovered antibiotics appear at very low frequencies of 10^{-7} per random actinomycete tested, since more common antibiotics appear at high frequencies and hide the least common ones [41].

Despite these challenges, recent research has proven that there are undiscovered sources of potential antibiotic natural products from new microbial ecological niches [16], such as lakes, rivers, marine environments, animals' gastrointestinal tract, extreme environments like the polar sea ice [42], ice cores [43], hydrothermal vents [44], sub-seafloor sites [45] or caves [46].

The oceans cover more than 70% of the earth's surface, representing an enormous resource for the discovery of potential natural products with interesting therapeutic properties. By using remotely operated vehicles, exploration of marine environments is possible at high depths without provoking any damages. Only in the mid-1970s, marine exploration for drug discovery was initiated and during the decade from 1977 to 1987, 2,500 new metabolites were reported from an array of marine organisms [7].

More recently, Blunt *et al.* (2014) [47] reviewed the marine natural product literature for 2012 (382 articles) and found 1,241 new marine natural products, corresponding to an increase of 8% in comparison with 2011. These studies show the richness of the marine environment in terms of bioactive compounds, many of them not found in terrestrial environment. Marine natural products present a great diversity, belonging to several different classes of compounds. The major focus of research is the marine microorganisms, presenting a rich array of novel and bioactive compounds. Green, brown and red algae, sponges, cnidarians, bryozoans, molluscs, tunicates, echinoderms and mangroves are the other groups of marine organisms from which are isolated many natural products [47]. Nonetheless, though there are several bioactive natural compounds isolated from marine organisms, the majority remains in the pharmaceutical pre-clinical trials and very few proceed to be marketed as pharmaceutical products [17].

Currently, there are seven FDA and/or European Medicines Agency (EMEA)-approved drugs originated from the marine environment on the market, mostly from macroorganisms; four of these compounds are used in the cancer treatment, namely cytarabine (Cytosar-U®, Depocyt®), trabectedin (Yondelis®), eribulin mesylate (Halaven®) and brentuximab (Adcetris®); vidarabine (Vira-A®) has anti-viral activity and is only used in Europe; ziconotide (Prialt®) is used in the relief of chronic pain; and omega-3-acid ethyl esters (Lovaza®) is used for the treatment of hypertriglyceridemia [48].

While searching for new antimicrobials from marine organisms, Plaza *et al.* [49] identified a methanol extract of the chrysophyte alga *Chrysophaeum taylori* which strongly inhibited the growth of methicillin-resistant-*Staphy- lococcus aureus* (MRSA) and vancomycin-resistant *Enterococcus faecium* (VRE). After isolating eight new compounds called chrysophaentin A-H from the extract, researchers assessed their antimicrobial activity and concluded that chrysophaentin A was the most potent antibiotic with MICs of 1 to 3 μg/ml against MRSA and VRE. Moreover, they showed that chrysophaentin A was able to inhibit the GTPase activity of FtsZ *in vitro*, a cytoskeletal protein essential for bacterial cell division which is highly conserved within all bacterial species [49].

Actinomycete strains were isolated from sediments and sponges collected in the Trondheim Fjord in Norway and their putative production of bioactive secondary metabolites was assessed by screening for genes involved in polyketide and nonribosomal peptide antibiotic synthesis using polymerase chain reaction (PCR) and by screening for the production of antibacterial and antifungal compounds from the fermentation extracts. The strongest antibacterial agent to be found was thiopeptide TP-1161, produced by Gram-positive *Nocardiopsis* sp. bacterium, in particular against *Enterococcus faecalis* strains (MICs 0.25-1.0 μg/ml) [50].

Two new sulfated sterols, geodisterol-3-*O*-sulfate and 29-demethylgeodisterol -3-*O*-sulfate were identified in a marine sponge from the genus *Topsentia*. They increased the activity of the antifungal agent fluconazole by reversing the efflux pump-mediated fluconazole resistance in *Candida albicans* [51].

Extracted from the marine sponge *Neopetrosia proxima*, found at the west coast of Puerto Rico, neopetrosiamine A displayed *in vitro* inhibitory activity against *Mycobacterium tuberculosis* $H_{37}Rv$ (MIC = 7.5 μg/ml) and *Plasmodium falciparum* (IC_{50} = 2.3 μM) (IC50 is the half maximal inhibitory concentration). Likewise, the marine compound showed strong inhibitory activity against three cancer cell lines, while exhibiting no significant cytotoxicity against Vero cells [52].

In 2010, Cheng and colleagues [53] reported three new diterpenoids from the soft coral *Sinularia gyrosa* termed as gyrosanols A-C. Compounds A and B showed antiviral activity against human cytomegalovirus (HCMV) with IC_{50} of 2.6 and 3.7 μM, respectively [53].

Additionally, plants have also been regarded as alternative sources for antimicrobial molecules. Like animals, plants possess defense mechanisms for combating pathogen infection, such as apoptosis of infected tissues, production of antimicrobial peptides, hydrogen peroxide or iron chelators [54]. Moreover, plants have developed a limited range of genomically encoded receptors to recognize specific pathogens and produce a great diversity of secondary metabolites with antimicrobial properties. Although plants produce over 100,000 small-molecule compounds with weaker antimicrobial activity than common antibiotics produced by bacteria and fungi, the interest in these sources hasn't faded, which is clearly demonstrated by a vast literature in this area of research [16].

Plants natural products serve as plant defense mechanisms against insects, herbivores and microorganisms [55]; while others are responsible for odor, flavor and pigmentation. The difficulty in finding antibiotics from plants with elevated and selective toxicity, potent and broad-spectrum antimicrobial activity may be due to the fact that plants only produce some compounds when a pathogen is attacking them, which leads to the conclusion that it may be an energetically costly action [54].

As it was previously mentioned, there is a great biodiversity among the plant kingdom and medicines based on the use of plants have been used for millennia to treat various maladies. More importantly, drugs used for the treatment of parasitic infections have been developed from plants, like artemisinin, an antimalarial agent, obtained from *Artemisia annua* [56].

The extensive range of plant-derived compounds has been reviewed in some articles [57, 58] in terms of their antimicrobial activity. The phytochemical diversity of compounds includes phenols and polyphenols (simple phenols, phenolic acids, quinones, flavonoids, tannins, coumarins), terpenoids and essential oils, alkaloids, lectins and polypeptides.

Phenolic compounds from medicinal herbs and dietary plants possess a range of bioactivities, such as antibacterial, antifungal, antiviral, antimutagenic and anti - inflammatory. In addition, according to clinical evidence, phenolic phytochemicals could be applied in chemotherapy and cancer prevention [59]. Jayaraman and colleagues [60] tested the *in vitro* antimicrobial activity of seven antibiotics in combination with six phytochemicals (protocatechuic acid, gallic acid, ellagic acid, rutin, berberine and myricentin) against five strains of *Pseudomonas aeruginosa* by assessing the MICs and performing a time-kill assay. These compounds had the ability to enhance the susceptibility of the pathogen to various antibiotics and decrease the toxicity of the drugs when used in combination [60]. By combining plant natural products with antibiotics, the antibiotic dosage could be reduced and the antibiotic adverse reactions would be, therefore, decreased.

Licarin A is a neolignan isolated from the hexane extract of *Aristolachia taliscana* roots, a Mexican plant used in traditional medicine to treat cough and snake bites. This compound was the most active of three isolated neolignans against four mono-resistant variants and twelve clinical isolates of *Mycobacterium tuberculosis*, with MICs ranging from 3.12 to 12.5 μg/ml [61]. These results suggest the potential bioactivity of neolignans against the tuberculosis agent, which has become more of a worldwide concern lately, with one third of the world's population being currently infected with the mycobacterium [62].

Quinones are organic compounds consisting of aromatic rings with two ketone substitutions and ubiquitous in nature [63]. These compounds have the ability to irreversibly complex with nucleophilic amino acids in proteins, which many times lead to their inactivation and loss of function. Besides, they also provide a source of stable free radicals [64]. Cell wall polypeptides, adhesins and membrane-bound enzymes are the probable targets. Anthraquinone is an example of a bacteriostatic agent against several pathogenic bacteria, like *Bacillus anthracis, Corynebacterium pseudodiphthericum* and *Pseudomonas aeruginosa* and bactericidal against *Burkholderia pseudomallei* [65].

Flavonoids are phenolic structures ubiquitous in photosynthetic cells, usually found in fruit, vegetables, nuts, seeds, stems, flowers, tea, wine, propolis and honey [64]. These compounds act as chemoattractants for insect pollinators and as signals molecules for beneficial rhizosphere microbiota. More importantly, flavonoids function as plant antimicrobial defense products [66]. Their antibacterial properties may come from the ability to form complexes with both extracellular and soluble proteins and with bacterial membranes, which leads to increase of membrane permeability and disruption. Eight flavonoids isolated from the roots of *Eriosema chinense* were reported to show relevant growth inhibition of *M. tuberculosis* H37Ra with **MICs** ranging from 12.5 to 50 μg/ml [67]. Likewise, seven flavonoids found in the methanolic extract of leaves of *Leiothrix spiralis* Ruhland showed promising antimicrobial activity against *Escherichia coli*, *P. aeruginosa* and four *Candida* species, including *C. albicans* [68].

Tannins are a group of water-soluble oligomeric and polymeric polyphenolic compounds, with significant astringent properties. Being present in the whole plant [69], tannins are widely used in various industries, namely leather and food.

Their antibacterial activity is attributable to the inactivation of bacterial adhesins, enzymes, cell envelope and transport proteins [70]. Engels and colleagues [71] reported the inhibitory activity of gallotannin-rich plant extracts isolated from mango kernels (*Mangifera indica* L.) against Gram-negative and Gram-positive bacteria, being the latter more susceptible to gallotannins antimicrobial activity, due to their affinity for iron. Knowing their spectrum and mode of action, purified gallotannins could be used as preservatives in food products and are compatible with the simultaneous use of starter cultures, because lactic acid bacteria are resistant to them [71].

Coumarins are a group of aromatic benzopyrones consisting of fused benzene and alpha pyrone rings. They are used as antithrombotic and anti-inflammatory compounds [72]. Three coumarins isolated from *Fatoua pilosa* as antitubercular compounds; isobavachalcone was the most inhibitory with MIC of 18 μg/ml, while scopoletin and (E)-1-2,4-dihydroxy-3-(3-methylbut-2-enyl) phenyl - 3-(2,--dimethyl-8-hydroxy-2H-benzopyran-6-yl) prop-2-en-1-one showed moderate activity; MICs of 42 and 30 μg/ml, respectively [73].

Miyasaki and colleagues [74] isolated several compounds from plant extracts which previously have shown to inhibit a multidrug-resistant clinical *Acinetobacter baumannii*. The most active compound was norwogonin with MIC of 128 μg/ml. Though its development as a systemic therapeutic may be limited due to its high MIC, further studies are necessary to establish the prophylactic and therapeutic potential of norwogonin for infections by multidrug-resistant bacteria [74].

Terpenes are one of the largest and most diverse groups of secondary metabolites, consisting of five carbon isoprene structural units linked in various configurations. The major classes consist of diterpenes, triterpenes, tetraterpenes, hemiterpenes and sesquiterpenes. When the compounds contain additional elements, usually oxygen, they are designated terpenoids.

Parasitic flatworms from the genus *Schistosoma* cause schistosomiasis, a neglected tropical disease which leads to a chronic and debilitating disease, one of the leading causes of morbidity and mortality in endemic countries [75]. Nerolidol is a sesquiterpene present in the essential oils of several plants. Tested concentrations of 31.2 and 62.5 μM reduced the *Schistosoma mansoni* adult worm motor activity and caused the death of all male and female schistosomes, respectively. Moreover, morphological changes on the tegument of worms were observed, demonstrating that nerolidol can be a promising compound for the development of antischistosomal natural agents [76].

Essential oils are also secondary metabolites of aromatic plants, being "essential" since they represent the most important part of the plant. In general, essential oils are produced using steam distillation and being very concentrated while simple plant extracts often engage the use of solvents, like acetone, ethanol or hexane, for extraction. Additionally, essential oils are highly complex volatile mixtures with 20 to 60 components in different concentrations, the most frequently found being terpenes and aromatic compounds [77]. Throughout the years, several essential oils have been reviewed as anti-inflammatory, sedative, digestive, antibacterial, antiviral, antifungal, antioxidant and cytotoxic agents [31, 78]. Due to their chemical and structural variance, essential oils could represent a unique group of putative new drug compounds.

Researchers reported the antimicrobial activity of essential oils from the leaves of *Cinnamomum pubescens* Kochummen and the whole plant of *Etlingera elatior* against MRSA, *Bacillus subtilis*, *P. aeruginosa* and *Salmonella choleraesuis*, being MRSA the most susceptible bacteria to *C. pubescens* Kochummen [79].

Lemongrass (*Cymbopogon citratus*) essential oil in vapor phase is a potent inhibitor of *C. albicans* growth, leading to deleterious morphological changes in cellular structures and cell surface alterations, when compared to lemongrass essential oil in liquid phase. Both mentha (*Mentha piperita*) and eucalyptus (*Eucalyptus globulus*) exhibited antifungal activity, but less than lemongrass. The alterations in the fungi surface leads to less adhesion and, therefore, less virulence and infectiousness. Vapor phase major advantage is in not requiring direct contact and having an easier application [80].

Essential oil from *Chenopodium ambrosioides* was reported to present *in vitro* antileishmanial activity against *Leishmania donovani*, causing an irreversible inhibition of promastigotes growth after a treatment with 100 or 10 µg/ml for 1 or 24 h, respectively [81].

Relatively to *Trypanosoma cruzi,* essential oils of *Cinnamomum verum, Myrocarpus frondosus* and *Eugenia uniflora* showed the highest inhibitory activity, with IC_{50} values of 24.13, 60.87 and 70 µg/ml, respectively. *C. verum* was effective on the three developmental forms of the parasite (epimastigotes, trypomastigotes and amastigotes) and on its *in vitro* differentiation [82].

In general, plant antimicrobial compounds are classified as constitutive (phytoanticipins) [83] when they are already in the plant tissue or as inducible (phytoalexins) [84, 85] when they are produced after pathogen attack [86]. However, there is a great difficulty in classifying them, since the same compound could be constitutive in some plant organs, but inducible in others.

According to González-Lamothe and colleagues [87], phytoalexins are "antimicrobial compounds which require *de novo* expression of the enzymes involved in their biosynthetic pathways after elicitation. Their production requires transcriptional and/or translational activity in the plant once the pathogen has been detected. The induced response mechanism also involves the trafficking and secretion of antimicrobial compounds to the infection site".

Both types present several different mechanisms of action; phytoanticipin avenacin, saponin from oat roots, can form complexes with sterols in the fungal membrane, which leads to membrane permeability [88]; phytoalexin scopoletin from the tobacco plants acts as a scavenger of reactive oxygen species excessively produced after triggering the hypersensitive response, *i.e.*, cells surrounding the site of infection are signaled to undergo apoptosis, when the plant is under attack [89, 90].

Interestingly, at the present time, no plant natural product has been approved for systemic use to combat bacterial infections, maybe because their spectrum of activity is very slim and/or the mode of action is nonspecific [87]. Isolation of specific active components from extracts has been hard, since all components are structurally related with varying degrees of bioactivity and/or cytotoxicity. Moreover, the isolated compounds could lead to gain of microbial resistance mechanisms, since they would be in single form and more easily expelled or degraded. Hence, the combined effects of the mixture of natural products found in plants might be necessary to obtain a synergistic and more efficient antibacterial activity [87].

Regarding natural plant products contribution to enhance the response against pathogenic microorganisms affecting humans, presently, there are two potential applications of phytoanticipins/phytoalexins for therapeutic use. Plant products called "antibiotic potentiators" could allow antibiotics to recover some of their lost therapeutic applications and "virulence attenuators" could support the host immune system to adequately respond to the pathogen invasion.

In order to defend themselves against antimicrobial agents, bacteria have developed and evolved several defense mechanisms. Efflux pumps are one of the mechanisms, being membrane translocases which extrude structurally unrelated toxins from the bacterial cell.

Plant natural products act as inhibitors of bacterial efflux pumps, responsible for intrinsic resistance against many toxic molecules. Thus, these pump-inhibitors can be used in combination with conventional antibiotics, potentiating their effect, *i.e.*, antibiotic potentiators. For example, green tea extracts possess the flavonoid catechin, a putative efflux pump inhibitor, which abolished tetracycline resistance in staphylococcal isolates expressing TetK, an efflux pump [91]. Catechin also potentiated the activity of β-lactam antibiotics by binding to the cell wall and sensitizing *Staphylococcus aureus* to low osmotic pressure [92].

The medicinal plants from genus *Berberis*, used in Native American traditional medicine, produce an antibacterial alkaloid called berberine. This hydrophobic cation accumulates in cells driven by the membrane potential and is a DNA intercalator [93], active on several microorganisms with a target on RNA polymerase, gyrase and topoisomerase IV and on nucleic acid [94]. However berberine shows relatively weak antibacterial properties because it is readily extruded by bacterial efflux pump NorA. NorA is a single-component pump of *S. aureus* part of the major facilitator superfamily (MFS) transporters which confers resistance to quinolones and antiseptics. Stermitz *et al.* [95] reported the action of a NorA efflux pump inhibitor, 5'-methoxyhydnocarpin (5'-MHC). 5'-MHC has no antibacterial effect on its own; however, the level of accumulation of berberine in the staphylococcal cells was strongly increased in the presence of 5'-MHC, indicating that this compound effectively disabled the bacterial resistance mechanism against the antimicrobial agent [95]. Years later, it was reported that covalently linking berberine to INF_{55}, an efflux-pump inhibitor, results in an effective antimicrobial hybrid molecule which showed good efficacy in a *Caenorhabditis elegans* model of enterococcal infection by showing good anti-infective activity [96].

Instead of directly killing pathogenic bacteria, virulence attenuators act by preventing the production of virulence factors, allowing the non-pathogenic and beneficial bacterial communities to remain unaffected. Also, less selective pressure would be exerted so that there would be no development of resistance, as it happens with commonly used antibiotics [97].

The adaptability of pathogens to the host environment could be prevented by blocking *quorum sensing* systems responsible for the synchronization of the infection process. Rasmussen *et al.* [98] designed a screening for *quorum sensing* inhibitory plant extracts by fusing reported genes to *quorum sensing*-controlled promoters. These authors reported garlic extract as one of the most potent *quorum sensing* inhibitors, able to reduce *Pseudomonas aeruginosa* biofilm tolerance to tobramycin treatment. Moreover, microarray-derived transcriptional analyses further confirmed the repression of *quorum sensing*-controlled gene expression when exposed to garlic extracts [98].

CONCLUDING REMARKS

Overall, natural product research continues to be a relevant foundation for antimicrobial drug discovery. Being secondary metabolites, natural products lack an apparent function in growth or even development on the producing organism. Instead, natural products have been studied due to their biological effect on other organisms, in particular against pathogenic microorganisms which cause problematic infectious diseases and are an actual challenge for human medicine.

Although there is still academic research about the antimicrobial properties of natural compounds, the pharmaceutical industry shows decreased interest in such area. Hence, in order to avoid the dangerous consequences of the emergence of infectious diseases, cooperation between academic and pharmaceutical researchers should be considered to find viable alternatives to currently used antibiotics.

CONFLICT OF INTEREST

The authors confirm that this chapter contents have no conflict of interest.

ACKNOWLEDGEMENTS

Declared none.

REFERENCES

[1] Sarker SD, Nahar L. An introduction to natural products isolation. In: Sarker SD, Nahar L, Eds. Natural Products Isolation. Totowa, NJ: Humana Press 2012; pp. 1-25.
[http://dx.doi.org/10.1007/978-1-61779-624-1_1]

[2] Cragg GM, Newman DJ. Natural product sources of drugs: Plants, microbes, marine organisms, and animals. In: Taylor JB, Triggle DJ, Eds. Comprehensive Medicinal Chemistry II. Elsevier 2007; pp. 355-403.
[http://dx.doi.org/10.1016/B0-08-045044-X/00011-0]

[3] Hanson JR. The classes of natural product and their isolation. Natural products: The secondary metabolites. Cambridge: Royal Society of Chemistry 2003; pp. 1-34.
[http://dx.doi.org/10.1039/9781847551535-00001]

[4] Ríos JL, Recio MC. Medicinal plants and antimicrobial activity. J Ethnopharmacol 2005; 100(1-2): 80-4.
[http://dx.doi.org/10.1016/j.jep.2005.04.025] [PMID: 15964727]

[5] Kingston DG. Modern natural products drug discovery and its relevance to biodiversity conservation. J Nat Prod 2011; 74(3): 496-511.
[http://dx.doi.org/10.1021/np100550t] [PMID: 21138324]

[6] Newman DJ, Cragg GM. Natural products as sources of new drugs over the 30 years from 1981 to 2010. J Nat Prod 2012; 75(3): 311-35.
[http://dx.doi.org/10.1021/np200906s] [PMID: 22316239]

[7] Cragg GM, Newman DJ. Natural products: A continuing source of novel drug leads. Biochimica et Biophysica Acta (BBA) -. General Subjects 2013; 1830: 3670-95.
[http://dx.doi.org/10.1016/j.bbagen.2013.02.008]

[8] Chin Y-W, Balunas MJ, Chai HB, Kinghorn AD. Drug discovery from natural sources. AAPS J 2006; 8(2): E239-53.
[http://dx.doi.org/10.1007/BF02854894] [PMID: 16796374]

[9] Cragg GM, Katz F, Newman DJ, Rosenthal J. The impact of the United Nations Convention on Biological Diversity on natural products research. Nat Prod Rep 2012; 29(12): 1407-23. [http://dx.doi.org/10.1039/c2np20091k] [PMID: 23037777]

[10] Ganesan A. The impact of natural products upon modern drug discovery. Curr Opin Chem Biol 2008; 12(3): 306-17. [http://dx.doi.org/10.1016/j.cbpa.2008.03.016] [PMID: 18423384]

[11] Gray AI, Igoli JO, Edrada-Ebel R. Natural products isolation in modern drug discovery programs. In: Sarker SD, Nahar L, Eds. Natural products isolation. Totowa, NJ: Humana Press 2012; pp. 515-34. [http://dx.doi.org/10.1007/978-1-61779-624-1_20]

[12] Singh SB, Barrett JF. Empirical antibacterial drug discovery--foundation in natural products. Biochem Pharmacol 2006; 71(7): 1006-15. [http://dx.doi.org/10.1016/j.bcp.2005.12.016] [PMID: 16412984]

[13] Koehn FE, Carter GT. The evolving role of natural products in drug discovery. Nat Rev Drug Discov 2005; 4(3): 206-20. [http://dx.doi.org/10.1038/nrd1657] [PMID: 15729362]

[14] Li JW, Vederas JC. Drug discovery and natural products: end of an era or an endless frontier? Science 2009; 325(5937): 161-5. [http://dx.doi.org/10.1126/science.1168243] [PMID: 19589993]

[15] Rishton GM. Natural products as a robust source of new drugs and drug leads: past successes and present day issues. Am J Cardiol 2008; 101(10A): 43D-9D. [http://dx.doi.org/10.1016/j.amjcard.2008.02.007] [PMID: 18474274]

[16] Taylor PW. Alternative natural sources for a new generation of antibacterial agents. Int J Antimicrob Agents 2013; 42(3): 195-201. [http://dx.doi.org/10.1016/j.ijantimicag.2013.05.004] [PMID: 23796893]

[17] Martins A, Vieira H, Gaspar H, Santos S. Marketed marine natural products in the pharmaceutical and cosmeceutical industries: tips for success. Mar Drugs 2014; 12(2): 1066-101. [http://dx.doi.org/10.3390/md12021066] [PMID: 24549205]

[18] Brady SF. Construction of soil environmental DNA cosmid libraries and screening for clones that produce biologically active small molecules. Nat Protoc 2007; 2(5): 1297-305. [http://dx.doi.org/10.1038/nprot.2007.195] [PMID: 17546026]

[19] Li X, Qin L. Metagenomics-based drug discovery and marine microbial diversity. Trends Biotechnol 2005; 23(11): 539-43. [http://dx.doi.org/10.1016/j.tibtech.2005.08.006] [PMID: 16154653]

[20] Kunst F, Ogasawara N, Moszer I, *et al.* The complete genome sequence of the gram-positive bacterium *Bacillus subtilis*. Nature 1997; 390(6657): 249-56. [http://dx.doi.org/10.1038/36786] [PMID: 9384377]

[21] Straight PD, Fischbach MA, Walsh CT, Rudner DZ, Kolter R. A singular enzymatic megacomplex from *Bacillus subtilis*. Proc Natl Acad Sci USA 2007; 104(1): 305-10. [http://dx.doi.org/10.1073/pnas.0609073103] [PMID: 17190806]

[22] Butcher RA, Schroeder FC, Fischbach MA, *et al.* The identification of bacillaene, the product of the PksX megacomplex in *Bacillus subtilis*. Proc Natl Acad Sci USA 2007; 104(5): 1506-9. [http://dx.doi.org/10.1073/pnas.0610503104] [PMID: 17234808]

[23] Challis GL. Mining microbial genomes for new natural products and biosynthetic pathways. Microbiology 2008; 154(Pt 6): 1555-69. [http://dx.doi.org/10.1099/mic.0.2008/018523-0] [PMID: 18524911]

[24] Lautru S, Deeth RJ, Bailey LM, Challis GL. Discovery of a new peptide natural product by Streptomyces coelicolor genome mining. Nat Chem Biol 2005; 1(5): 265-9. [http://dx.doi.org/10.1038/nchembio731] [PMID: 16408055]

[25] Bentley SD, Chater KF, Cerdeño-Tárraga A-M, *et al.* Complete genome sequence of the model actinomycete Streptomyces coelicolor A3(2). Nature 2002; 417(6885): 141-7. [http://dx.doi.org/10.1038/417141a] [PMID: 12000953]

[26] Piel J. Combinatorial biosynthesis in symbiotic bacteria. Nat Chem Biol 2006; 2(12): 661-2. [http://dx.doi.org/10.1038/nchembio1206-661] [PMID: 17108981]

[27] de Bruijn I, de Kock MJ, Yang M, de Waard P, van Beek TA, Raaijmakers JM. Genome-based discovery, structure prediction and functional analysis of cyclic lipopeptide antibiotics in Pseudomonas species. Mol Microbiol 2007; 63(2): 417-28. [http://dx.doi.org/10.1111/j.1365-2958.2006.05525.x] [PMID: 17241198]

[28] Othman M, Loh HS, Wiart C, Khoo TJ, Lim KH, Ting KN. Optimal methods for evaluating antimicrobial activities from plant extracts. J Microbiol Methods 2011; 84(2): 161-6. [http://dx.doi.org/10.1016/j.mimet.2010.11.008] [PMID: 21094190]

[29] Das K, Tiwari RK, Shrivastava DK. Techniques for evaluation of medicinal plant products as antimicrobial agent: Current methods and future trends. J Med Plants Res 2010; 4: 104-11.

[30] Hammer KA, Carson CF, Riley TV. Antimicrobial activity of essential oils and other plant extracts. J Appl Microbiol 1999; 86(6): 985-90. [http://dx.doi.org/10.1046/j.1365-2672.1999.00780.x] [PMID: 10438227]

[31] Burt S. Essential oils: their antibacterial properties and potential applications in foods--a review. Int J Food Microbiol 2004; 94(3): 223-53. [http://dx.doi.org/10.1016/j.ijfoodmicro.2004.03.022] [PMID: 15246235]

[32] Seow YX, Yeo CR, Chung HL, Yuk H-G. Plant essential oils as active antimicrobial agents. Crit Rev Food Sci Nutr 2014; 54(5): 625-44. [http://dx.doi.org/10.1080/10408398.2011.599504] [PMID: 24261536]

[33] Mann CM, Markham JL. A new method for determining the minimum inhibitory concentration of essential oils. J Appl Microbiol 1998; 84(4): 538-44. [http://dx.doi.org/10.1046/j.1365-2672.1998.00379.x] [PMID: 9633651]

[34] McMichael AJ. Environmental and social influences on emerging infectious diseases: past, present and future. Philos Trans R Soc Lond B Biol Sci 2004; 359(1447): 1049-58. [http://dx.doi.org/10.1098/rstb.2004.1480] [PMID: 15306389]

[35] Aminov RI. A brief history of the antibiotic era: lessons learned and challenges for the future. Front Microbiol 2010; 1: 134.

[http://dx.doi.org/10.3389/fmicb.2010.00134] [PMID: 21687759]

[36] Davies J, Davies D. Origins and evolution of antibiotic resistance. Microbiol Mol Biol Rev 2010; 74(3): 417-33.
[http://dx.doi.org/10.1128/MMBR.00016-10] [PMID: 20805405]

[37] Furuya EY, Lowy FD. Antimicrobial-resistant bacteria in the community setting. Nat Rev Microbiol 2006; 4(1): 36-45.
[http://dx.doi.org/10.1038/nrmicro1325] [PMID: 16357859]

[38] Shah PM. The need for new therapeutic agents: what is the pipeline? Clin Microbiol Infect 2005; 11 (Suppl. 3): 36-42.
[http://dx.doi.org/10.1111/j.1469-0691.2005.01141.x] [PMID: 15900655]

[39] Talbot GH, Bradley J, Edwards JE Jr, Gilbert D, Scheld M, Bartlett JG. Antimicrobial availability task force of the infectious diseases society of america. Bad bugs need drugs: an update on the development pipeline from the antimicrobial availability task force of the infectious diseases society of america. Clin Infect Dis 2006; 42(5): 657-68.
[http://dx.doi.org/10.1086/499819] [PMID: 16447111]

[40] Watve MG, Tickoo R, Jog MM, Bhole BD. How many antibiotics are produced by the genus Streptomyces? Arch Microbiol 2001; 176(5): 386-90.
[http://dx.doi.org/10.1007/s002030100345] [PMID: 11702082]

[41] Baltz R. Antimicrobials from Actinomycetes: Back to the Future. Microbe 2007; 2: 125-31.

[42] Murray AE, Grzymski JJ. Diversity and genomics of Antarctic marine micro-organisms. Philos Trans R Soc Lond B Biol Sci 2007; 362(1488): 2259-71.
[http://dx.doi.org/10.1098/rstb.2006.1944] [PMID: 17553778]

[43] Antony R, Krishnan KP, Laluraj CM, *et al.* Diversity and physiology of culturable bacteria associated with a coastal Antarctic ice core. Microbiol Res 2012; 167(6): 372-80.
[http://dx.doi.org/10.1016/j.micres.2012.03.003] [PMID: 22537873]

[44] Hensley SA, Jung J-H, Park C-S, Holden JF. Thermococcus paralvinellae sp. nov. and *Thermococcus cleftensis sp.* nov. of hyperthermophilic heterotrophs from deep-sea hydrothermal vents. Int J Syst Evol Microbiol 2014; 64(Pt 11): 3655-9.
[http://dx.doi.org/10.1099/ijs.0.066100-0] [PMID: 25082851]

[45] Forschner-Dancause S, LaPlante K, Smith DC, Rowley DC. Going deep for drug discovery: an ocean to bedside approach to explore sub-seafloor microbes for the next generation of antibiotics. Med Health R I 2012; 95(9): 292-3.
[PMID: 23094427]

[46] Ortiz M, Legatzki A, Neilson JW, *et al.* Making a living while starving in the dark: metagenomic insights into the energy dynamics of a carbonate cave. ISME J 2014; 8(2): 478-91.
[http://dx.doi.org/10.1038/ismej.2013.159] [PMID: 24030597]

[47] Blunt JW, Copp BR, Keyzers RA, Munro MH, Prinsep MR. Marine natural products. Nat Prod Rep 2014; 31(2): 160-258.
[http://dx.doi.org/10.1039/c3np70117d] [PMID: 24389707]

[48] Mayer AMS. 2014. The Global Marine Pharmaceuticals Pipeline. Available from:

http://marinepharmacology.midwestern.edu/index.htm. , [2014 Dec 23.];

[49] Plaza A, Keffer JL, Bifulco G, Lloyd JR, Bewley CA. Chrysophaentins A–H. Antibacterial bisdiarylbutene macrocycles that inhibit the bacterial cell division protein FtsZ. J Am Chem Soc 2010; 132: 9069-77.
[http://dx.doi.org/10.1021/ja102100h] [PMID: 20536175]

[50] Engelhardt K, Degnes KF, Kemmler M, *et al.* Production of a new thiopeptide antibiotic, TP-1161, by a marine Nocardiopsis species. Appl Environ Microbiol 2010; 76(15): 4969-76.
[http://dx.doi.org/10.1128/AEM.00741-10] [PMID: 20562278]

[51] Digirolamo JA, Li X-C, Jacob MR, Clark AM, Ferreira D. Reversal of fluconazole resistance by sulfated sterols from the marine sponge Topsentia sp. J Nat Prod 2009; 72(8): 1524-8.
[http://dx.doi.org/10.1021/np900177m] [PMID: 19653640]

[52] Wei X, Nieves K, Rodríguez AD. Neopetrosiamine A, biologically active bis-piperidine alkaloid from the Caribbean sea sponge Neopetrosia proxima. Bioorg Med Chem Lett 2010; 20(19): 5905-8.
[http://dx.doi.org/10.1016/j.bmcl.2010.07.084] [PMID: 20727745]

[53] Cheng S-Y, Chuang C-T, Wang S-K, *et al.* Antiviral and anti-inflammatory diterpenoids from the soft coral Sinularia gyrosa. J Nat Prod 2010; 73(6): 1184-7.
[http://dx.doi.org/10.1021/np100185a] [PMID: 20499851]

[54] Lewis K, Ausubel FM. Prospects for plant-derived antibacterials. Nat Biotechnol 2006; 24(12): 1504-7.
[http://dx.doi.org/10.1038/nbt1206-1504] [PMID: 17160050]

[55] Jones JD, Dangl JL. The plant immune system. Nature 2006; 444(7117): 323-9.
[http://dx.doi.org/10.1038/nature05286] [PMID: 17108957]

[56] Paddon CJ, Westfall PJ, Pitera DJ, *et al.* High-level semi-synthetic production of the potent antimalarial artemisinin. Nature 2013; 496(7446): 528-32.
[http://dx.doi.org/10.1038/nature12051] [PMID: 23575629]

[57] Cowan MM. Plant products as antimicrobial agents. Clin Microbiol Rev 1999; 12(4): 564-82.
[PMID: 10515903]

[58] Saleem M, Nazir M, Ali MS, *et al.* Antimicrobial natural products: an update on future antibiotic drug candidates. Nat Prod Rep 2010; 27(2): 238-54.
[http://dx.doi.org/10.1039/B916096E] [PMID: 20111803]

[59] Huang W-Y, Cai Y-Z, Zhang Y. Natural phenolic compounds from medicinal herbs and dietary plants: potential use for cancer prevention. Nutr Cancer 2010; 62(1): 1-20.
[http://dx.doi.org/10.1080/01635580903191585] [PMID: 20043255]

[60] Jayaraman P, Sakharkar MK, Lim CS, Tang TH, Sakharkar KR. Activity and interactions of antibiotic and phytochemical combinations against *Pseudomonas aeruginosa in vitro*. Int J Biol Sci 2010; 6(6): 556-68.
[http://dx.doi.org/10.7150/ijbs.6.556] [PMID: 20941374]

[61] León-Díaz R, Meckes M, Said-Fernández S, *et al.* Antimycobacterial neolignans isolated from Aristolochia taliscana. Mem Inst Oswaldo Cruz 2010; 105(1): 45-51.
[http://dx.doi.org/10.1590/S0074-02762010000100006] [PMID: 20209328]

[62] Daglia M. Polyphenols as antimicrobial agents. Curr Opin Biotechnol 2012; 23(2): 174-81. [http://dx.doi.org/10.1016/j.copbio.2011.08.007] [PMID: 21925860]

[63] Upadhyay A, Upadhyaya I, Kollanoor-Johny A, Venkitanarayanan K. Combating pathogenic microorganisms using plant-derived antimicrobials: a minireview of the mechanistic basis. BioMed Res Int 2014; 2014: 761741. [http://dx.doi.org/10.1155/2014/761741] [PMID: 25298964]

[64] Savoia D. Plant-derived antimicrobial compounds: alternatives to antibiotics. Future Microbiol 2012; 7(8): 979-90. [http://dx.doi.org/10.2217/fmb.12.68] [PMID: 22913356]

[65] Kazmi MH, Malik A, Hameed S, Akhtar N, Noor Ali S. An anthraquinone derivative from Cassia italica. Phytochemistry 1994; 36: 761-3. [http://dx.doi.org/10.1016/S0031-9422(00)89812-X]

[66] Dixon RA, Steele CL. Flavonoids and isoflavonoids - a gold mine for metabolic engineering. Trends Plant Sci 1999; 4(10): 394-400. [http://dx.doi.org/10.1016/S1360-1385(99)01471-5] [PMID: 10498963]

[67] García A, Bocanegra-García V, Palma-Nicolás JP, Rivera G. Recent advances in antitubercular natural products. Eur J Med Chem 2012; 49: 1-23. [http://dx.doi.org/10.1016/j.ejmech.2011.12.029] [PMID: 22280816]

[68] Araújo MG de F, Hilário F, Nogueira LG, Vilegas W, Santos LC, Bauab TM. Chemical constituents of the methanolic extract of leaves of Leiothrix spiralis Ruhland and their antimicrobial activity. Molecules 2011; 16(12): 10479-90. [http://dx.doi.org/10.3390/molecules161210479] [PMID: 22179427]

[69] Scalbert A. Antimicrobial properties of tannins. Phytochemistry 1991; 30: 3875-83. [http://dx.doi.org/10.1016/0031-9422(91)83426-L]

[70] Haslam E. Natural polyphenols (vegetable tannins) as drugs: possible modes of action. J Nat Prod 1996; 59(2): 205-15. [http://dx.doi.org/10.1021/np960040+] [PMID: 8991956]

[71] Engels C, Schieber A, Gänzle MG. Inhibitory spectra and modes of antimicrobial action of gallotannins from mango kernels (Mangifera indica L.). Appl Environ Microbiol 2011; 77(7): 2215-23. [http://dx.doi.org/10.1128/AEM.02521-10] [PMID: 21317249]

[72] Namba T, Morita O, Huang S-L, Goshima K, Hattori M, Kakiuchi N. Studies on cardio-active crude drugs; I. Effect of coumarins on cultured myocardial cells. Planta Med 1988; 54(4): 277-82. [http://dx.doi.org/10.1055/s-2006-962432] [PMID: 3222369]

[73] Chiang C-C, Cheng M-J, Peng C-F, Huang H-Y, Chen I-S. A novel dimeric coumarin analog and antimycobacterial constituents from Fatoua pilosa. Chem Biodivers 2010; 7(7): 1728-36. [http://dx.doi.org/10.1002/cbdv.200900326] [PMID: 20658660]

[74] Miyasaki Y, Rabenstein JD, Rhea J, *et al.* Isolation and characterization of antimicrobial compounds in plant extracts against multidrug-resistant Acinetobacter baumannii. PLoS One 2013; 8(4): e61594. [http://dx.doi.org/10.1371/journal.pone.0061594] [PMID: 23630600]

[75] Steinmann P, Keiser J, Bos R, Tanner M, Utzinger J. Schistosomiasis and water resources development: systematic review, meta-analysis, and estimates of people at risk. Lancet Infect Dis 2006; 6(7): 411-25.
[http://dx.doi.org/10.1016/S1473-3099(06)70521-7] [PMID: 16790382]

[76] Silva MP, Oliveira GL, de Carvalho RB, *et al.* Antischistosomal activity of the terpene nerolidol. Molecules 2014; 19(3): 3793-803.
[http://dx.doi.org/10.3390/molecules19033793] [PMID: 24662089]

[77] Yap PS, Yiap BC, Ping HC, Lim SH. Essential oils, a new horizon in combating bacterial antibiotic resistance. Open Microbiol J 2014; 8: 6-14.
[http://dx.doi.org/10.2174/1874285801408010006] [PMID: 24627729]

[78] Bakkali F, Averbeck S, Averbeck D, Idaomar M. Biological effects of essential oils--a review. Food Chem Toxicol 2008; 46(2): 446-75.
[http://dx.doi.org/10.1016/j.fct.2007.09.106] [PMID: 17996351]

[79] Abdelwahab SI, Zaman FQ, Mariod AA, Yaacob M, Abdelmageed AH, Khamis S. Chemical composition, antioxidant and antibacterial properties of the essential oils of Etlingera elatior and Cinnamomum pubescens Kochummen. J Sci Food Agric 2010; 90(15): 2682-8.
[http://dx.doi.org/10.1002/jsfa.4140] [PMID: 20945508]

[80] Tyagi AK, Malik A. Liquid and vapour-phase antifungal activities of selected essential oils against Candida albicans: microscopic observations and chemical characterization of Cymbopogon citratus. BMC Complement Altern Med 2010; 10: 65.
[http://dx.doi.org/10.1186/1472-6882-10-65] [PMID: 21067604]

[81] Monzote L, García M, Montalvo AM, Scull R, Miranda M, Abreu J. In vitro activity of an essential oil against Leishmania donovani. Phytother Res 2007; 21(11): 1055-8.
[http://dx.doi.org/10.1002/ptr.2210] [PMID: 17724767]

[82] Azeredo CM, Santos TG, Maia BH, Soares MJ. In vitro biological evaluation of eight different essential oils against Trypanosoma cruzi, with emphasis on Cinnamomum verum essential oil. BMC Complement Altern Med 2014; 14: 309.
[http://dx.doi.org/10.1186/1472-6882-14-309] [PMID: 25148924]

[83] Wittstock U, Gershenzon J. Constitutive plant toxins and their role in defense against herbivores and pathogens. Curr Opin Plant Biol 2002; 5(4): 300-7.
[http://dx.doi.org/10.1016/S1369-5266(02)00264-9] [PMID: 12179963]

[84] Hammerschmidt R. PHYTOALEXINS: What Have We Learned After 60 Years? Annu Rev Phytopathol 1999; 37: 285-306.
[http://dx.doi.org/10.1146/annurev.phyto.37.1.285] [PMID: 11701825]

[85] Smith CJ. Tansley Review No. 86 Accumulation of phytoalexins: defence mechanism and stimulus response system. New Phytol 1996; 132: 1-45.
[http://dx.doi.org/10.1111/j.1469-8137.1996.tb04506.x]

[86] Bonanomi G, Vinale F, Scala F. The role of natural products in plant-microbe interactions. In: Osbourn AE, Lanzotti V, Eds. Plant-derived natural products. New York, NY: Springer US 2009; pp. 301-20.

[http://dx.doi.org/10.1007/978-0-387-85498-4_14]

[87] González-Lamothe R, Mitchell G, Gattuso M, Diarra MS, Malouin F, Bouarab K. Plant antimicrobial agents and their effects on plant and human pathogens. Int J Mol Sci 2009; 10(8): 3400-19. [http://dx.doi.org/10.3390/ijms10083400] [PMID: 20111686]

[88] Morrissey JP, Osbourn AE. Fungal resistance to plant antibiotics as a mechanism of pathogenesis. Microbiol Mol Biol Rev 1999; 63(3): 708-24. [PMID: 10477313]

[89] Chong J, Baltz R, Fritig B, Saindrenan P. An early salicylic acid-, pathogen- and elicitor-inducible tobacco glucosyltransferase: role in compartmentalization of phenolics and H2O2 metabolism. FEBS Lett 1999; 458(2): 204-8. [http://dx.doi.org/10.1016/S0014-5793(99)01154-0] [PMID: 10481066]

[90] Costet L, Fritig B, Kauffmann S. Scopoletin expression in elicitor-treated and tobacco mosaic virus-infected tobacco plants. Physiol Plant 2002; 115(2): 228-35. [http://dx.doi.org/10.1034/j.1399-3054.2002.1150208.x] [PMID: 12060240]

[91] Sudano Roccaro A, Blanco AR, Giuliano F, Rusciano D, Enea V. Epigallocatechin-gallate enhances the activity of tetracycline in staphylococci by inhibiting its efflux from bacterial cells. Antimicrob Agents Chemother 2004; 48(6): 1968-73. [http://dx.doi.org/10.1128/AAC.48.6.1968-1973.2004] [PMID: 15155186]

[92] Zhao W-H, Hu Z-Q, Okubo S, Hara Y, Shimamura T. Mechanism of synergy between epigallocatechin gallate and beta-lactams against methicillin-resistant *Staphylococcus aureus*. Antimicrob Agents Chemother 2001; 45(6): 1737-42. [http://dx.doi.org/10.1128/AAC.45.6.1737-1742.2001] [PMID: 11353619]

[93] Iwasa K, Moriyasu M, Yamori T, Turuo T, Lee D-U, Wiegrebe W. In vitro cytotoxicity of the protoberberine-type alkaloids. J Nat Prod 2001; 64(7): 896-8. [http://dx.doi.org/10.1021/np000554f] [PMID: 11473418]

[94] Yi Z-B, Yan Yu , Liang Y-Z, Bao Zeng . Evaluation of the antimicrobial mode of berberine by LC/ESI-MS combined with principal component analysis. J Pharm Biomed Anal 2007; 44(1): 301-4. [http://dx.doi.org/10.1016/j.jpba.2007.02.018] [PMID: 17383137]

[95] Stermitz FR, Lorenz P, Tawara JN, Zenewicz LA, Lewis K. Synergy in a medicinal plant: antimicrobial action of berberine potentiated by 5′-methoxyhydnocarpin, a multidrug pump inhibitor. Proc Natl Acad Sci USA 2000; 97(4): 1433-7. [http://dx.doi.org/10.1073/pnas.030540597] [PMID: 10677479]

[96] Ball AR, Casadei G, Samosorn S, *et al.* Conjugating berberine to a multidrug efflux pump inhibitor creates an effective antimicrobial. ACS Chem Biol 2006; 1(9): 594-600. [http://dx.doi.org/10.1021/cb600238x] [PMID: 17168555]

[97] Bjarnsholt T, Givskov M. Quorum-sensing blockade as a strategy for enhancing host defences against bacterial pathogens. Philos Trans R Soc Lond B Biol Sci 2007; 362(1483): 1213-22. [http://dx.doi.org/10.1098/rstb.2007.2046] [PMID: 17360273]

[98] Rasmussen TB, Bjarnsholt T, Skindersoe ME, *et al.* Screening for quorum-sensing inhibitors (QSI) by use of a novel genetic system, the QSI selector. J Bacteriol 2005; 187(5): 1799-814.

[http://dx.doi.org/10.1128/JB.187.5.1799-1814.2005] [PMID: 15716452]

CHAPTER 7

Bacteriocins

Sara Correia Santos[1], Manuela Oliveira[2], Teresa Semedo Lemsaddek[2,*]

[1] *Eco-evolutionary Genetics Group, Instituto Gulbenkian de Ciência, Rua da Quinta Grande 6, 2780-156 Oeiras, Portugal*

[2] *Centro de Investigação Interdisciplinar em Sanidade Animal, Faculdade de Medicina Veterinária, Universidade de Lisboa, Avenida da Universidade Técnica, 1300-477 Lisboa, Portugal*

Abstract: Bacteriocins are antimicrobial peptides produced by numerous bacteria, which may present narrow or broad host range. These ribosomally synthesized antimicrobial peptides are considered a successful strategy in maintaining equilibrium within a bacterial ecosystem. These compounds kill other bacteria by several mechanisms, including the modification of membrane permeability and depolarization of membrane ion gradients, the degradation of nucleic acids or cell walls. In literature, the term bacteriocin is usually restricted to peptides produced by Gram-positive bacteria, while in Gram-negative bacteria, mainly enterobacteria, the toxins are called either colicins (*i.e.* antibiotic proteins targeting *Escherichia coli*) or microcins. Many bacteriocins are produced by food-grade lactic acid bacteria, a phenomenon which offers the possibility for preventing the development of specific bacterial species in food. This can be particularly useful in preservation or food safety applications, but also has implications for the development of desirable flora in fermented food. In this sense, bacteriocins can be used to confer a protection and at the same time help processors extend shelf-life after product manufacture

Keywords: Anticompetitors, Antimicrobial-peptides, Archeosins, Bacteriocins, Bacteriolytic enzymes, Biopreservation, BLIS, Bovamine, Colicins, Detection, Enterocin, Food-preservatives, Genetic-regulation, Halocin, Hemolysin, Immunity, LAB, Lacticin, Lantibiotics, Lysostaphin, Mersacidin, Microcins, Mode of action, Nisaplin, Nisin, Novasin, Pediocin, Pre-peptide, Secretion, Spot-on-lawn, Streptolysin, Toxin.

[*] **Corresponding author Teresa Semedo-Lemsaddek:** Centro de Investigação Interdisciplinar em Sanidade Animal (CIISA), Faculdade de Medicina Veterinária da Universidade de Lisboa, Lisboa, Portugal; E-mail: tlemsaddek@fmv.ulisboa.pt

Manuela Oliveira and Isa Serrano (Eds)

1. INTRODUCTION

The use of antimicrobial peptides is one of the most widespread strategies against microbes. These substances are produced from insects to plants and even humans. Antimicrobial peptides can be fast acting and usually present broad-spectrum activity, which diminishes the possibility of resistance developing in target species. Production of these peptides can also be found in Bacteria and Archaea. Bacteria produce an enormous diversity of antimicrobial agents including antibiotics, metabolic byproducts, such as lactic acids produced by lactic acid bacteria (LAB), lytic agents such as lysozymes and numerous types of protein exotoxins, such as bacteriocins.

Bacteriocins are ribosomally synthesized, small, heat-stable peptides that upon release extracellularly can interfere with the growth of other bacteria. Bacteriocins can present a narrow (same species) or broad (across genera) target spectrum. Producer-organisms are immune to their own bacteriocins through the production of specific immunity proteins [1 - 3].

The term *bacteriocins* was first defined by Tagg *et al.* (1976) [4] as proteinaceous compounds able to kill closely related bacteria. While true for the majority of bacteriocins, for some, it could also be true for bacteria confined to the same ecological niche [5]. In 1877, Pasteur and Joubert reported an inhibitory activity against the anthrax bacilli associated with bacteria isolated from urine, probably *Escherichia coli*. The first studies on bacteriocins were performed by Gratia and Fredericq (1925) on Gram-negative bacteria, especially colicins produced by *E. coli* [6]. By this time, bacteriocins were defined based on colicins properties such as being plasmid-encoded, large domain-structured proteins, having bactericidal activity through specific receptors and being lethal SOS-inducible biosynthesis (Fredericq, 1946) [7]. However, few Gram-positive bacteriocins fit this classical colicin definition. The main differences are further explored in this chapter (see "Diversity of Bacteriocins in Gram-positive Bacteria"). Shortly, they include relatively broad activity spectra; producer cell self-protection (immunity); and absence of SOS- inducibility. During the last decades, studies in Gram-positive bacteria bacteriocins have dominated bacteriocin-related literature, with special focus on Lactic Acid Bacteria (LAB), driven by commercial interests [2].

There are two main features that distinguish bacteriocins from antibiotics: (i) bacteriocins are ribosomally synthesized, while antibiotics are not; and (ii) bacteriocins have a relatively narrow killing spectrum, while most antibiotics present broad killing ranges. The bacteriocin family includes a number of proteins that vary in size, microbial target, mode of action, release and immunity mechanisms, and can be divided into two main groups, those produced by Gram-negative or by Gram-positive bacteria.

Bacteriocins are found in almost every bacterial species examined to date, and within a species tens or even hundreds of different kinds of bacteriocins are present. According to Klaenhammer, 99% of all bacteria may produce at least one bacteriocin and the only reason we have not found more is that we haven't looked for them [8]. This diversity and abundance of bacteriocins suggest an important role for these antimicrobials agents. These toxins have been found in all major lineages of Bacteria, and more recently, have been described as universally produced by some members of the Archaea (see "Bacteriocins from Archaea"). In this chapter we summarize literature that has helped unravel the mysteries of bacteriocins since their discovery; explore diversity, genetic regulation and mode of action and applications in food safety and human and livestock health.

2. BACTERIOCINS IN GRAM-NEGATIVE BACTERIA

The first description of bacteriocin-mediated inhibition was reported in 1925 by Gratia and Fredericq in antagonistic strains of *E. coli* [6]. These compounds were originally designated as colicins to reflect the producing organism. Bacteriocins produced by Gram-negative bacteria are diverse. Just in *E.coli* over 30 different bacteriocin forms have been identified.

Levels of bacteriocin production in Gram-negative bacteria, such as *E. coli*, *Salmonella enterica*, *Hafnia alvei*, *Citrobacter freundii*, *Klebsiella oxytoca*, *Klebsiella pneumoniae*, and *Enterobacter cloacae* range from 3 to 26% of environmental isolates. Colicins, produced by *E. coli*, were found in 30–50% of human strains. Much higher production levels have been found in some other bacteria, such as *Pseudomonas aeruginosa*, in which 90% or more of both environmental and clinical isolates produce bacteriocins [9].

2.1. Colicins

Since their discovery, colicins have been extensively studied and now serve as model systems for research on the mechanisms of bacteriocin structure/function, genetic organization, ecology, and evolution. Colicins are high molecular weight proteins that kill target cells through a variety of mechanisms [10], being able to inhibit macromolecular synthesis without arrest of respiration (colicin E1 and K), cause DNA breakdown (colicin E2) and stop protein synthesis (colicin E3) [11].

2.2. Microcins

The microcins represent the model for modified peptide antibiotics produced by Gram-negative bacteria. Microcins are a peculiar class of gene-encoded low molecular mass antibacterial peptides, secreted by enterobacteria. They contribute to the regulation of microbial competition within the intestinal microbiota. The genetic systems involved in microcin biosynthesis present a conserved organization. Similar to bacteriocins of Gram-positive bacteria, microcins exert potent antibacterial closely related bacterial strains. They display a great structural diversity among the few representatives well characterized until now, that makes difficult to describe the subclasses of microcin [12, 13].

2.3. Genetic regulation of Colicins and Colicin-like bacteriocins

Colicins are usually encoded on one of two types of colicinogenic plasmids. Nevertheless, plasmid carriage is not a requirement. Type 1 plasmids are small (6–10 kb), present in numerous copies per cell and are mobilized in the presence of a conjugative plasmid. Type 2 are single copy plasmids of about 40 kb, which carry numerous genes in addition to those encoding colicin activity, which can be transferred by conjugation [10].

The general colicin operon organization is common to almost all colicins. The first gene (*cxa*) encodes for colicin activity, followed downstream by the gene encoding an immunity protein (designated either *cxi* or *immX*). The immunity gene encodes a specific immunity protein to the producer cell that binds to and inactivates the toxin. The last gene, encodes a protein involved in colicin release through lysis of the producer cell *(cxl)* [10].

Differently from bacteriocins of Gram-positive bacteria, which are self-regulated and have specifically dedicated transport mechanisms, colicins and other bacteriocins of Gram-negative bacteria are dependent on the regulatory pathways of the host cell, mainly the SOS system. Under conditions of stress, a small proportion of the bacteriocin-producing cells are induced. The producing cells commit 'suicide', resulting in the release of the toxin molecules killing neighbor sensitive cells. The released proteins bind to specific cell surface receptors on target cells and are translocated into the cell *via* a set of transport proteins. These specific targeting and transport mechanisms ensure a narrow killing range. For a more detailed revision on colicin translocation please read the report by Klean Thous, 2010 [14]. The toxin then kills the target cell by one of several mechanisms: (i) channel formation in the cytoplasmic membrane; (ii) cellular DNA degradation; (iii) protein biosynthesis inhibition by cleaving RNA; or (iv) murein and lipopolysaccharide biosynthesis inhibition by interference with lipid carrier regeneration. It is assumed that such targeted killing results in increased resources for the closest relatives of the producing cells, which are immune to the toxin by encoding the specific immunity protein [10].

Colicins have three functionally distinct domains:

- Receptor recognition domains (comprises about 50% of the protein and is involved in the recognition of specific cell surface receptors),
- Protein translocation domain (the N-terminal domain, <25% of the protein, is responsible for translocation of the protein into the target cell),
- Killing and immunity domain (a short sequence involved in immunity protein binding).

Most Gram-negative bacteriocins have an identical domain structure. Regulation of colicins seems to be dependent not only on DNA damaging agents, but also on catabolic repression, stringent response, cell phase, anaerobiosis and for some, the expression of other colicins. Other factors trigger low or no response, such as osmolarity, nutrient depletion, catabolite repression, heavy metals, pH and temperature [10].

2.4. Mode of Action

Colicins action is divided in three steps. First, the toxin binds to a specific cell surface receptor on the outer membrane; secondly it is translocated through the cell envelope by either the Tol or the TonB machinery to its target, which are the inner membrane for ionophore colicins and the cytoplasm for nuclease colicins. The third step corresponds to target-cell killing [10, 14].

Bacteriocins from Gram-positive bacteria, particularly LAB, and microcins from enterobacteria, mostly *E. coli*, share a number of common characteristics. They have a low molecular mass (<10 kDa); their secretion involves ABC-type transporters; the producer strain has a specific immunity mechanism that protects it from being killed by the newly synthesized antibacterial molecule; and both types of peptide inhibit the growth of phylogenetically related species. However, microcins form a much more restricted group with highly diversified structures and mechanisms of action. This is why microcins have also diverse mechanisms of action, targeting either the inner membrane of target bacteria, or enzymes involved in DNA or RNA structure and synthesis or in protein synthesis. A common feature shared by microcins is the use of specific receptors designed by all bacteria for essential nutrient uptake to enter the bacteria more efficiently. The microcins describe to date share a common 'Trojan horse' strategy for killing other bacteria. Either the microcin mimics essential molecules through its three-dimensional structure allowing recognition by naturally specific receptors designed by bacteria for the uptake of essential elements. Or, the microcin is secreted as a harmless molecule that is allowed to enter the susceptible bacteria and once inside, the inactive microcin is further transformed by the host bacterium itself into a fatal toxin. Such strategies differ immensely from those adopted by Gram-positive bacteriocins [13].

3. BACTERIOCINS OF GRAM-POSITIVE BACTERIA

Bacteriocins of Gram-positive bacteria are as abundant and even more diverse as those found in Gram-negative bacteria. They differ in two fundamental ways. First, bacteriocin production is not necessarily the lethal event for Gram-positive as it is for Gram-negative bacteria. This critical difference is due to the transport mechanisms of Gram-positive bacteria. Some developed a bacteriocin-specific transport system, while others use the sec-dependent export pathway. Secondly, Gram-positive bacteria have established bacteriocin-specific regulation, whereas bacteriocins of Gram-negative bacteria rely solely on host regulatory networks [10, 11].

Most Gram-positive bacteria, especially LAB, are food-grade organisms already widespread in the food industry, with applications in food preservation. LAB produce a great variety of compounds with the ability of inhibiting growth of both competing microbiota and food spoilage microorganisms. Such metabolic byproducts include organic acids, antibiotics, lytic agents, toxins, bacteriolytic enzymes (such as lysozyme), bacteriophages and other byproducts, such as hydrogen peroxide and diacetyl [4, 15, 16]. Since LAB and their metabolites have been present (naturally or as starter cultures) in foods with no adverse effects, they continue to be a preferred source for food-use bacteriocins, either by use of purified compounds or growth extracts. There have been many studies demonstrating that LAB used in cheese manufacture to convert lactose to lactic acid also produce bacteriocins that can influence the composition of the complex cheese microbiota and potentially inhibit adventitious spoilage or pathogenic bacteria [1]. LAB bacteriocins are effective against key Gram-positive pathogens of importance in foodborne illness, including *Listeria monocytogenes*, a common pathogenic bacterium difficult to control in food products [16].

3.1. Classification

Criteria on bacteriocin definition evolved with increasing knowledge. The term was first used by Jacob *et al.* (1953) [17] as protein-like substances, the biosynthesis of which would imply death for the producing organism. The genes coding for bacteriocins and host immunity would be located in plasmids and the release of this proteins would be lethal to its producer. Meanwhile, around 1976 the definition was broadened since it was verified that most bacteriocins produced by Gram-positive do not follow this definition. They possess a specific defense mechanism that protects the producer strain upon release of the bacteriocin [4, 18].

Furthermore, the term bacteriocin should not be used for inhibitory substances that are not fully characterized. For these cases the term bacteriocin-like inhibitory substance (BLIS) should be applied [2]. The designation of a specific bacteriocin should only occur after the isolation, purification and sequencing of the peptide(s) and corresponding structural genes [2].

Bacteriocins with minor conservative differences in the amino acid sequences of their pro-peptide components, with no significant change to (i) their secondary structure, (ii) their activity spectra, or (iii) the cross-specificity of their producer strain immunity, should be referred to as natural variants. This is the case of nisins Z, Q and U, which are natural variants of the first described nisin A [2].

In 1993, Klaenhammer [5], proposed four major classes of Gram-positive bacteria bacteriocins:

- Class I, Lanthionine-containing bacteriocins or Lantibiotics;
- Class II, Small heat-stable, non-lanthionine containing membrane-active peptides;
- Class III, Large heat-labile proteins;
- Class IV, Complex bacteriocins, composed of protein plus one or more chemical moieties (lipid, carbohydrate) required for activity.

In 2005, Cotter *et al.* [1] proposed a very different scheme of classifying LAB bacteriocins. This classification is much simpler, dividing bacteriocins in two categories, the Class I Lantibiotics and the Class II non-lanthionine-containing bacteriocins. The former Class III of large heat-labile murein hydrolases was renamed as bacteriolysins, and class IV (the lipid- or carbohydrate-containing bacteriocins) was removed. Klaenhammer Class II subgroups IIa (listeria-active peptides) and IIb (two-peptide bacteriocins) was retained and the class V (cyclic peptides) proposed by Kemperman *et al.* (2003) [19] was reassigned as class IIc. Class IId was proposed to be a repository for all of the remaining linear non-lanthionine-containing bacteriocins.

Later, in 2007 Heng *et al.* suggested some alterations to the Cotter *et al.* scheme (Fig. **1**). For more detailed information regarding classification of subtypes specificities read the reviews by Cotter *et al.* (2005) and Heng *et al.* (2007) [1, 2].

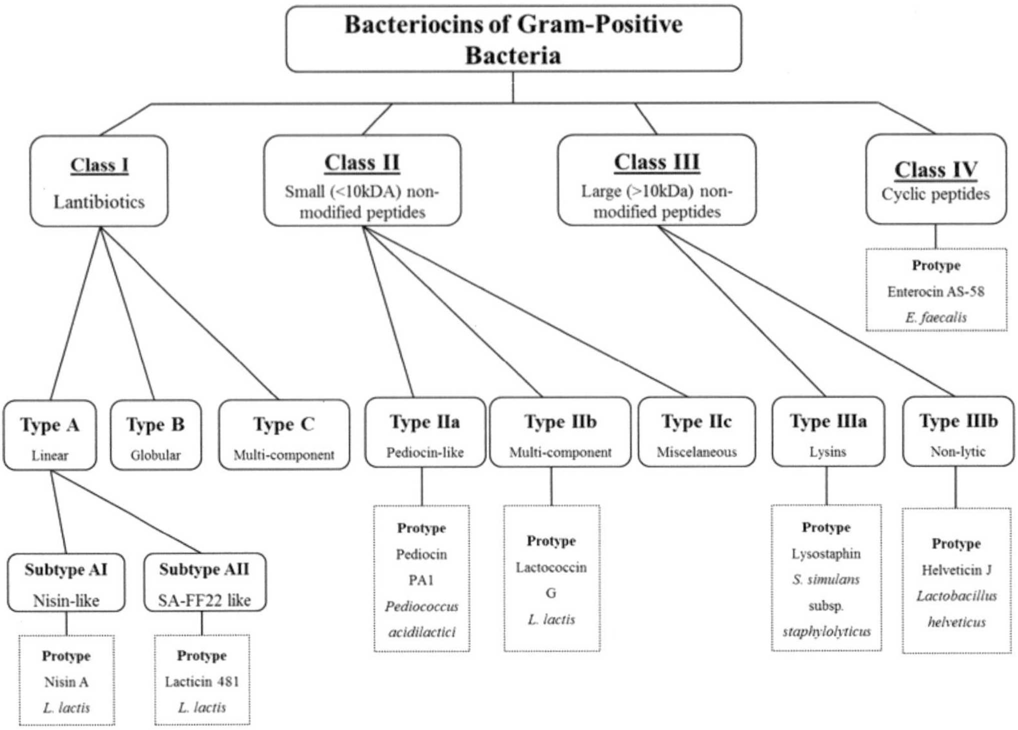

Fig. (1). Classification scheme for Gram-positive bacteria bacteriocins. [Adapted from Heng *et al.,* 2007 [2]].

3.1.1. Class I: The Lanthionine-Containing (Lantibiotics) Bacteriocins

Lantibiotics are a diverse group of small (19-38 amino acids) bactericidal peptides containing a non-genetically encoded amino acids lanthionine (Lan) and/or β-methyllanthionine (MeLan). These unusual residues form covalent bridges between amino acids, which results in internal 'rings' and gives lantibiotics their characteristic structural features. Furthermore, lantibiotics can contain other unusual residues that result from post-translational modification, including the substitution of d-alanine for l-serine [1]. All of the lantibiotics currently described are thought to be produced as ribosomally synthesized precursor peptides, which then undergo a series of post- translational modification reactions to produce the unusual amino acids described above as components of the biologically active peptides.

To date, lantibiotics are known to be produced only by Gram-positive bacteria and act predominantly, if not exclusively, on Gram-positive targets.

The elongated amphiphilic cationic lantibiotics (for example, nisin) are active through the formation of pores, leading to the dissipation of membrane potential and the efflux of small metabolites from sensitive cells. By contrast, the globular lantibiotics (for example, mersacidin) act through enzyme inhibition. It has been shown that some bacteriocins, like nisin, possess both mechanisms and that other can function cooperatively [1].

They are divided into three subtypes according to their ring structures and their biological activities (Fig. **1**). Their complex structures make them a difficult group to subclassify. Type A represents elongated amphipathic structures; type B includes globular and more compact structures; and type C includes globular lantibiotics. The type A lantibiotics are further divided into subtypes AI and AII based on the size, charge and sequence of their leader peptides [2]. Among the LAB bacteriocins, the lantibiotic class is unique in their efficacy and established safety. Type A lantibiotics, such as lacticin, are small (2–5 kDa), elongated, screw-shaped proteins that contain positively charged molecules, which kill by membrane depolarization [10].

Type B lantibiotics, such as mersacidin, are smaller (about 2 kDa), globular in shape, and kill by interfering with cellular enzymatic reactions, such as cell wall synthesis [10].

3.1.2. Class II: The Unmodified Peptide Bacteriocins

The common non-lanthionine containing bacteriocins are small (<10 kDa) heat-stable peptides but, unlike lantibiotics, they are not subject to extensive post-translational modification. The majority of class II bacteriocins induce membrane permeabilization and the subsequent leakage of molecules from target bacteria. Their heterogeneous nature makes them difficult to classify. Two types are common to all classification systems: the class IIa pediocin-like or *Listeria*-active and the class IIb two-peptide bacteriocins. Pediocin-like bacteriocins have a narrow spectrum of activity but display a high specific activity against the food pathogen *Listeria monocytogenes*. They possess a conserved N-terminal sequence, a cationic "pediocin box" which contains the amino-acid sequence motif

(YGNGVXaaC) thought to facilitate nonspecific binding to the target surface. The C-terminal domains are less conserved and thought to determine non-listerial spectrum. Based on the C-terminus domain, three subdivisions were proposed (subclasses IIa, IIb and IIc) (Fig. **1**) [1, 10]. Like type A lantibiotics, class IIa bacteriocins act through the formation of pores in the cytoplasmic membrane. Examples include pediocin AcH, sakacin A, and leucocin A. Class IIb bacteriocins such as lacticin F and lactococcin G form pores, composed of two different proteins, in the membrane of their target cells. A third subgroup (IIc) consists of bacteriocins that are sec-dependent (such as acidocin B). Class III bacteriocins are large heat-labile proteins such as helveticins J and V and lactacin B. An additional proposed class (VI) requires lipid or carbohydrate moieties for activity. Little is known about the structure and function of this class. Examples include leuconocin S and lactocin 27 [10].

Classes I and II have been the focus of most probiotic research [10].

3.1.3. Class III: The Large (>10 kDa) Bacteriocins

Bacteriolysins (or class III bacteriocins) are large, heat-labile antimicrobial proteins. Their structure is characterized by different domains with functions for translocation, receptor binding and lethal activity. The mechanism of action is distinct from that of bacteriocins, as they function through the lysis of sensitive cells by catalyzing cell-wall hydrolysis. These proteins are also modular in structure and have a catalytic domain at the N-terminus that shows homology to endopeptidases, and a C-terminus that probably represents the target recognition site. Unlike the 'true' bacteriocins, they do not always have specific immunity genes that accompany bacteriocin structural genes, but might rely on modifications of the producer cell wall to impart resistance [1].

Bacteriocin research is progressing and it is not easy to formulate solid and natural classification schemes for all the existing bacteriocin and bacteriocin-like proteins. Especially when considering that most appear to have independent evolutionary origins and that there are a unimaginable variety of undescribed compounds.

3.2. Nisin: The Classic Case-Study

Nisin, the prototype lantibiotic type AI, is a LAB bacteriocin with commercial applications in food processing and fermentation [2, 5]. It is a classic example of a commercially successful naturally produced inhibitory agent. Produced by *Lactococcus lactis,* nisin is thought to have been first described in 1928 by Rogers and in 1933 by Whitehead. Later, in 1971, it was structurally characterized as a lanthionine-containing peptide (Gross, E., and J. L. Morell. 1971). Nisin was first marketed in England in 1953 and has been approved for use in over 48 countries. Its safety for food use was assessed by the Joint Food and Agriculture Organization/ World Health Organization Expert Committee on Food Additives in 1969. In 1983, this bacteriocin was added to the European food additive list as number E234 and, in 1988, it was approved by the US Food and Drug Agency (FDA) for use in pasteurized, processed cheese spreads [1].

Nisin and nisin-producing strains have long been applied in food preservation, especially in dairy products [20]. Nisin remains the most commercially important bacteriocin, although other bacteriocins have been characterized and developed for possible approval and use [16]. Remarkably it has been used in the food industry for decades without the appearance of significant bacterial resistance [2].

Nisin is not the only bacteriocin produced by *L. lactis*, these bacteria also produce non-lanthionine bacteriocins with a narrower spectrum of action. Diplococin is produced by *L. lactis* subsp. *cremoris* and has inhibitory action against other strains of *Lactococcus*. This bacteriocin does not induce lysis but instead inhibits nucleic acid synthesis. The genes encoding for this bacteriocin and its immunity are found to be in a conjugative plasmid. Another example of a bacteriocin produced by *L. lactis* subsp. *cremoris* is lactostrepcin. It has optimal activity at pH 4.6-5.0 and it binds to specific cell wall receptors, acting mainly at the membrane, destabilizing its integrity through loss of K^+, ATP and other low molecular weight compounds [8, 21].

Another example is lacticin 3147, a broad-spectrum bacteriocin produced by *Lactococcus lactis* subsp. *lactis*, which has been shown to be active against a range of food-borne bacteria that include methicillin-resistant *Staphylococcus aureus*, vancomycin - resistant *Enterococcus faecalis*, penicillin-resistant *Pneumococcus*, *Propionibacteriumacne* and *Streptococcus mutans* [22].

3.3. Genetic Regulation

Bacteriocins are usually synthesized as an inactive pre-peptide that includes an N-terminal leader sequence (called the double-glycine leader) whose function seems to be preventing the bacteriocin from being biologically active while inside the producer, and to provide the recognition signal for the transporter system [5]. Probably, in the case of lantibiotics, it also has a role in recognition by the modification machinery. This leader is usually cleaved during export by a dedicated bacteriocin-transport system or, less frequently, by the general secretion (Sec) pathway of the cell. A few bacteriocins seem to lack leader sequences [1].

Many, but not all, of the bacteriocins produced by Gram-positive bacteria seem to be encoded by plasmid borne genes. Some, like nisin, have been shown to be transposon associated. The bacteriocin associated genes of Gram-positive bacteria appear to be characteristically arranged in multigene operon-like structures. The genes that encode the structural pre-peptides are usually closely associated with genes that encode products involved in regulation, export, self-immunity and modification (in the case of lantibiotics) (Fig. **2**). The export of bacteriocins is usually achieved by a dedicated membrane-associated ATP-binding cassette (ABC) transporter that can also contain a proteolytic N-terminal domain belonging to the family of cysteine proteases that is responsible for cleavage of the leader peptide. In the case of some lantibiotics, cleavage might be due to a specific serine protease. For class II peptides, accessory proteins are thought to facilitate membrane translocation and/or leader peptide cleavage. There are several variations to this general scheme, which involve both N- and C-terminal or extracellular cleavage. Regulation of bacteriocin production and immunity is most frequently mediated through two-component signal-transduction systems, often as part of a quorum-sensing mechanism. In certain instances, more than one response regulator can be involved [1].

Bacteriocin producers have developed a protection system against their own bacteriocin, referred to as immunity. These systems enable a distinction between 'self' (producer) and 'non-self. Two main distinct systems are involved in class I and class II immunity, although there are some exceptions. As for many antibiotics, protection can be provided by a dedicated immunity protein and/or a specialized ABC-transporter system that pumps the bacteriocin from the producer membrane. Although immunity proteins rarely resemble one another, the general mechanism through which they function is probably similar, involving either the sequestration of the structural protein or an antagonistic competition for a receptor. Such immunity mechanisms are highly specific and usually do not provide protection against other bacteriocins. It has been shown that immunity to bacteriocins is brought about through the alteration of inter-peptide cross-bridges in peptidoglycan [1].

It has been occasionally observed that bacteriocinogenic LAB can lose their ability to produce bacteriocins. Some of these observations have been attributed to plasmid loss or to transposition-mediated inactivation. Recently, it has been shown that production of some bacteriocins can be transcriptionally regulated and that bacteriocins cannot be produced in the absence of an induction factor (IF). This phenomenon may also explain why loss of bacteriocin production is occasionally observed.

The production of some bacteriocins, like nisin, is self-inducible. The biosynthesis of lantibiotics and nonlantibiotics is usually regulated through two component regulatory systems (histidine protein kinase and a cytoplasmic response regulator). In this signal transduction pathway, the histidine-kinase protein auto-phosphorylates the conserved histidine residue in its intracellular domain, when it senses a certain concentration of bacteriocin in the environment. The phosphoryl group is subsequently transferred to the conserved aspartic acid residue on the response regulator domain and the resulting intramolecular change triggers the response regulator to activate the transcription of the regulated genes. These regulated genes include the structural gene, the export genes, the immunity genes, and in some cases, the regulatory genes themselves [23]. In contrast, most class II bacteriocins produce a bacteriocin-like peptide with no antimicrobial activity and use it as an induction factor (IF) to activate the transcription of the regulated genes. The IF is a small, heat-stable, cationic and hydrophobic peptide that is first synthesized as pre-peptide with a double glycine leader sequence. A dedicated

ABC-transporter specifically cleaves the leader peptide of IF concomitant with export of the mature peptide from the cell. The secreted IF acts as an external signal that triggers transcription of the genes involved in bacteriocin production [24].

The nature of the bacteriocins can be exceptionally diverse. Some can be encoded in conjugative or pheromone-responsive plasmids. The genetic transfer mechanisms, like conjugation, may explain the production of a variety of bacteriocins and the fact that multiple bacteriocins can be produced by a single strain; and that the same compound can be produced by different strains. On the other hand antibiotic resistance genes can also be present in these elements. In this way bacteriocin production can contribute as a competitive advantage for producers, especially allied with antibiotic resistance may enable them to establish themselves among a diverse microbiota, colonize and infect [25]. For the same reasons, bacteriocin production may also play an important role in food colonization, especially in dairy or meat fermented products [25].

3.3.1. Genetic regulation in Lantibiotics – Nisin

As mentioned before, nisin has been safely used in the food industry as a preservative for many decades. It is naturally produced by *Lactococcus lactis* and its biosynthesis pathway requires the expression of at least 11 gene products. This pathway is usually similar for most lantibiotics.

The nisin gene cluster includes genes for the prepeptide (nisA), enzymes for modifying amino acids (nisB, nisC), cleavage of the leader peptide (nisP), secretion (nisT), immunity (nisI, nisFEG), and regulation of expression (nisR, nisK) (Fig. **2**).

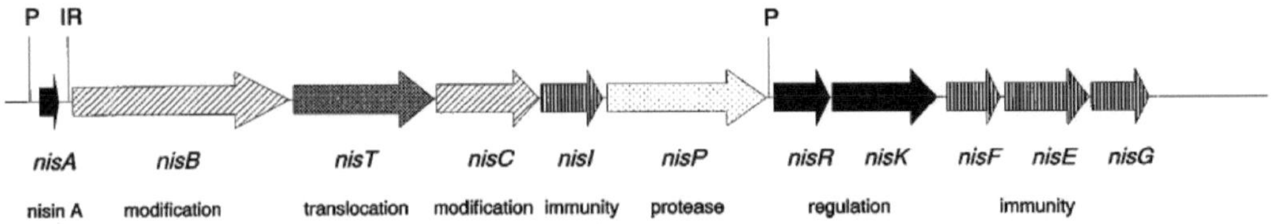

Fig. (2). Organization of nisin biosynthetic gene cluster. [Adapted from (Kuipers *et al.*, 1995) [23]].

These gene clusters are most often encoded on plasmids but are occasionally found on the chromosome. Several Gram-positive bacteriocins, including nisin, are located on transposons [10].

The precursor peptide, encoded by *nisA*, undergoes dehydration, by the proteins NisB and NisC, on particular Ser/Thr residues, some of which are then used to form specific thioether bonds (*i.e.*, Lan and MeLan) with Cys residues located (generally) further toward the C-terminus of the molecule. NisT is an ABC transporter responsible for export of the modified prepeptide, and NisP is a membrane-anchored protease able to cleave the leader peptide to release active nisin. NisI is involved in nisin immunity by an ill-defined mechanism. Although the majority of NisI appears to be localized within the cytoplasmic membrane of the producer cell, a significant amount is also secreted into the cytoplasm where it may bind to external nisin before it can aggregate at the cell surface. However, full immunity is not achieved without the presence of NisFEG. NisFEG is an ABC transport protein complex that contributes to nisin immunity by its concentration in direct contact with the cytoplasmic membrane.

NisR and NisK together form a two-component sensor-kinase/response-regulator element involved in the regulation of nisin biosynthesis, which usually occurs in late exponential growth phase. Interestingly, nisin up-regulates its own expression, since is the specific ligand recognized by the sensor NisK [23, 26]. NisK senses the presence of nisin in the medium and autophosphorylates. The phosphate group is transferred to NisR, which acts as a transcriptional activator, followed by mRNA synthesis and ribosomal synthesis of unmodified precursor nisin.

The precursor is then modified by the NisB and NisC. The fully modified precursor peptide is translocated across the membrane by an ABC transporter NisT. Once the modified precursor is extracellularly processed by NisP, it results in the release of active nisin. NisI, together with NisF, NisE, and NisG, protects the cell from the bacteriocidal action of nisin by an unknown mechanism [23]. The basic elements of the nisin biosysthesis pathway are conserved for all lantibiotics, with only minor variations [2].

Production of bacteriocins in Gram-positive bacteria is generally associated with the shift from log phase to stationary phase. For nisin, production begins during mid-log phase and increases to a maximum as the cells enter stationary phase. The regulation of expression is not cell cycle-dependent but culture density-dependent. Nisin A acts as a pheromone regulating its own expression, which is controlled by a two-component signal transduction system typical of many quorum-sensing systems, as mentioned before. The genes involved are nisR (the response regulator) and nisK (the sensor kinase) [10].

3.4. Mode of Action

Although a precise mechanism of action has not been elucidated for every lantibiotic, it is apparent that a docking molecule or target is involved, which has been established as LIPID II in several cases. The anionic lipids of the cytoplasmic membrane are the primary receptors for bacteriocins of LAB for initiation of pore formation. Conductivity and stability of pores induced by lantibiotics may be heightened by docking molecules (lipid II, the peptidoglycan precursor), while in the case of class II bacteriocins, receptors in the target membrane apparently act to determine specificity [16].

The activity of nisin is mediated by the binding of lipid II, which is also the target for therapeutic antibiotics such as vancomycin. A set of core amino acids at the N-terminal domain of nisin bind to an invariable region of lipid II thereby helping to avoid resistance like that observed in vancomycin-resistant enterococci (VRE).The binding of nisin to lipid II facilitates a dual mechanism of action involving the prevention of peptidoglycan synthesis and pore formation [1].

It is important to note that some bacteriocins of Gram-positive bacteria can also act as virulence factors. The enterococcal two-peptide lantibiotic cytolysin has activity against a broad spectrum of cell types, including a wide range of Gram-positive bacteria, eukaryotic cells such as human, bovine and horse erythrocytes, retinal cells, polymorphonuclear leukocytes and human intestinal epithelial cells, and is associated with acute, terminal outcomes in human infection. Furthermore, another non- lanthionine-containing peptide hemolysin, streptolysin S, is involved in invasive infection by some group A streptococci [1].

4. Bacteriocins of Archaea

Archaea also produce a distinct family of bacteriocin-like antimicrobials, known as archaeocins. The only characterized archeocins are the halocin family produced by halobacteria [11]. The first halocin discovered was S8, a short hydrophobic peptide of 36 amino acids. S8 is encoded on a mega-plasmid and is extremely stable, it can be desalted, boiled, subjected to organic solvents, and stored at 4ºC for extended periods without losing activity. Expression is growth stage-dependent. Archaeocins are produced as the cells enter stationary phase. When resources are limited, lysis of sensitive cells enriches the nutrient content of the environment. Although basal levels are present in low concentrations during exponential growth, there is an exponential increase in production during the transition to stationary phase. These are very stable proteins that can remain in the environment long enough to reduce competition during subsequent phases of nutrient flux. The mechanism of halocin action has been established only for halocin H6 but the immunity mechanism is unknown [10].

5. Detecting Bacteriocin Production

Generally, the methods for detecting bacteriocin production *in vitro* are agar culture-based. There are two main simple methods designated deferred and simultaneous antagonism [2]. However, there are some bacteriocins that appear to be produced only in liquid media, such as streptocins STH_1 and STH_2. After growth, the culture in liquid media is centrifuged and filtered in order to remove all cells. The solution with the secreted bacteriocins can be used to screen for sensitive strains by "spot-on-lawn" [27].

The simultaneous antagonism method is based on inhibition of growth of an indicator strain (sensitive) in agar plates in the presence of a producer strain. When the test and indicator bacteria are grown simultaneous on an agar surface, detection of bacteriocin production is dependent on the release of the inhibitory agent(s), which should occur relatively early in the growth of the test culture, before it is overgrown by the indicator strain. Indication of sensibility to a putative bacteriocin is observed by a halo on the indicator strains growth. The size of this inhibition halo is dependent on the diffusion and amount of compound produced.

It could also be dependent on the growth rate of the indicator strain, which in itself can be influenced by growth conditions such as growth medium, incubation time and temperature.

The deferred antagonism method is the most commonly used method for detecting bacteriocin production since it allows independent variation in incubation parameters (time, temperature and atmosphere) of the test and the indicator bacteria. In other words, this test allows screening of antagonistic activity as well as monitoring expression, optimization and inactivation of bacteriocins for their characterization. This test guarantees that optimal conditions for the test strain growth are used, since they may not coincide with optimal conditions for bacteriocin production. Even though, this is a straight forward test, it could present some challenges, especially when dealing with LAB, since they can produce other antagonistic compounds that could mimic bacteriocin activity [2, 28]. Using solid media, the producing bacteria is grown first under optimal conditions for growth and or bacteriocin production.

Both types of agar media screening do not distinguish the activities of bacteriocins from inhibition due to other types of agents such as bacteriophages, primary metabolites such as H_2O_2 and lactic acid, or non-ribosomally synthesized antibiotics, such as bacitracin. Neither can they discern inhibition through nutrient depletion, or combined activities from multiples bacteriocins and/or inhibitory agents.

For these cases, as aforementioned, the use of the acronym BLIS (bacteriocin-like inhibitory substance) is recommended, since it refers to yet uncharacterized inhibitory agents that appear to be "bacteriocin-like" in their activity [2, 4]. For example to screen for H_2O_2 inhibition, the test must be done in the presence of catalase; while screening for phage inhibition the "slam-transfer" method should be applied. In this method the producer strain is grown in an agar medium which is later inverted. Over this, a new layer of semi-solid agar is poured containing the sensitive strain. The bacteriocin is able to diffuse through the agar contrarily to phages [4, 27].

It is also important to screen the influence of the culture media on the bacteriocin production. To do this, a series of different agar-media should be used during the screening. This will allow determining the influence of agar pH and diffusion on bacteriocin production [27].

6. Promising Results on the Usage of Bacteriocins

6.1. In the Food Industry

Bacteriocins can be useful to protect food against contamination, or prevent the growth of specific pathogenic bacteria. Despite their effectiveness as foods preservatives, their use has become limited for various reasons, namely cost remains an issue, obstructing broader use of bacteriocins as food additives. As a result, there is a continued effort to find new and more effective bacteriocins, but also to improve existing bacteriocins to address both biologic and economic concerns [16]. Bacteriocins are now being explored for their potential utility in human and animal health applications, as well as agricultural uses [1].

The study on the inhibitory effect of bacteriocins intensified after Pasteur and Joubert, in 1877, have observed bacteria antagonizing *Bacillus anthracis* growth [4]. From then on, there was an increasing attempt to find non-pathogenic bacteria to fight disease. The excessive use of antibiotics in animal production and human medicine has resulted in effectiveness problems. This is one of the reasons why the research in bacteriocins has increased, in order to find molecules able to prevent growth of pathogenic bacteria such as *Listeria monocytogenes* and *Salmonella* sp. that could be safely used in the food industry. In 1964, the U. S. Food and Drug Administration (FDA) approved the use of nisin in food industry has a preservative, designating it as generally recognized as safe – GRAS [18].

Initially the study was mainly aimed at colicins, but eventually turned to bacteriocins produced by Gram-positive bacteria, especially LAB. Many of these food-grade organisms are widespread in the food industry to improve food preservation [18]. The application of bacteriocins in food biocontrol is mainly oriented in two alternative directions, the use of bacteriocin-producing bacteria or direct addition of bacteriocin.

Because LAB and their metabolites have been consumed in high quantities by countless generations of people with no adverse effects, they continue to be the preferred source for food-grade bacteriocins. Given the use of LAB as starter cultures in the manufacture of fermented foods, they are obvious candidates for genetic modification should industrial strains lack genes regulating synthesis of desirable bacteriocin(s). In addition to a sterling safety record and a public perception of LAB as health aids, many bacteriocins from these GRAS (generally regarded as safe) bacteria are effective against key Gram-positive pathogens of importance in foodborne illness, including *Listeria monocytogenes*, a resilient pathogenic bacterium common in the environment and difficult to control in foods [16].

Incorporation of bacteriocins into packaging films to control food spoilage and pathogenic organisms has been an area of active research for the last decade. Antimicrobial packaging film prevents microbial growth on food surface by direct contact of the package with the surface of foods, such as meats and cheese. For this reason, for it to work, the antimicrobial packaging film must contact the surface of the food so that bacteriocins can diffuse to the surface. The gradual release of bacteriocins from a packaging film to the food surface may have an advantage over dipping and spraying foods with bacteriocins [29].

The reason for the low number of bacteriocins currently used in food industry is related with several important criteria that should be taken into account when screening for bacteriocins to use in foods. It is important that the bacteriocin producing strain possesses a 'generally recognized as safe' (GRAS) status. Moreover, the bacteriocin should have a broad spectrum of inhibition that includes pathogens, or have activity against a particular pathogen with high specific activity. The bacteriocin should be heat stable and have no associated health risks. Finally, its inclusion in products should lead to beneficial effects such as improved safety, quality and flavor [1]. Bacteriocins have been shown to have potential in the biopreservation of meat, dairy products, canned food, fish, alcoholic beverages, salads, egg products, high-moisture bakery products, and fermented vegetables, either alone, in combination with other methods of preservation, or through their incorporation into packaging film/food surfaces [16].

Bacteriocins can be used as food preservatives in at least three different ways. In fermented food, they can be produced *in situ* by starter cultures; can also be added directly into foods in purified or semi-purified form, as the example of Nisaplin® and Novasin™ (for nisin); or as a fermentation product of a bacteriocin-producing strain, for example ALTA 2431, for pediocin PA-1 [30].

Bacteriocins can also be used to promote quality, rather than simply to prevent spoilage or safety problems. For example, bacteriocins can be used to control adventitious non-starter microbiota such as non-starter lactic acid bacteria (NSLAB) in cheese and wine that can cause major economic losses owing to calcium-d-lactate formation and slit defects in cheeses, and the production of detrimental compounds in wine. Bacteriocins can also promote quality by enhancing food fermentation. When applied during semi-hard and hard cheese manufacture, they can help control lysis of starter LAB, which results in the release of intracellular enzymes and ultimately accelerates ripening and may improve flavor [1].

Bacteriocins are an attractive option that could help to answer the consumer demands in regards to foods with a long shelf-life with minimal processing and reduced chemical preservatives. The *in situ* production of bacteriocins may increase the competitiveness of the producer strain in the food matrix and contribute to the prevention of spoilage. Bacteriocins are a good alternative to chemical food preservative since they are produced by food-grade organisms, they are usually heat stable and they can inhibit many of the primary pathogenic and spoilage organisms that cause problems in minimally processed foodstuffs. However, at present, only nisin and pediocin PA1/AcH have found widespread use in food [1].

6.2. In Human Health

As a consequence of the rising incidence of resistance to most antibiotics, numerous research programs have been implemented aiming to explore the potential role which naturally produced and genetically modified bacteriocins might have as replacements for such compounds [11].

Production of antimicrobials by a probiotic culture inside the human host could contribute to the inhibition of pathogenic bacteria in the gut. Since bacteriocins in food are degraded by the proteolytic enzymes of the stomach, probiotic bacteria might be ingested in a form that allows the *in vivo* production of the bacteriocin in the intestine [1].

Bacteriocins provide an alternative solution. Contrary to most antibiotics, they have relatively narrow spectrum of killing activity, and can be used to target specific bacterial pathogens. Given the diversity of bacteriocins produced in nature, it could be easy to find bacteriocins active against specific human pathogens [10].

Nisin and other members of the lantibiotic class have been investigated for their potential applications in medicine. Mota-Meira *et al.* (2000) [31] described MICs (minimal inhibitory concentration) of mutacin B-Ny266 and nisin A to be comparable to those of vancomycin and oxacillin against various bacterial pathogens.

These compounds were active against vancomycin- and oxacillin-resistant strains of *Helicobacter pylori* and *Neisseria* spp., making them potential candidates for treatment of infections caused by these bacteria.

Another example is lacticin 3147 that has shown *in vitro* activity against *Staphylococcus aureus* (including methicillin-resistant *S. aureus* MRSA), enterococci (including VRE, vancomycin resistant enterococci) and streptococci [22].

Nisin has also been investigated as a spermicidal contraceptive by Reddy *et al.* (2004) [32]. In this study vaginal administration of 1 mg of nisin in rabbits stopped sperm motility completely, having none of the treated animals become pregnant. *Streptococcus salivarius* K12 has been used as an oral probiotic strain releasing two lantibiotics (salivaricin A2 and salivaricin B) to antagonize the growth of *S. pyogenes* [33].

The normal vaginal microbiota is normally dominated by *Lactobacillus* spp., while the infected vagina is dominated by strains such as *Gardnerella vaginalis, Candida albicans, Propionibacterium, Lactobacillus, Streptococcus,*

Enterococcus, Bacteroides spp. L. acidophilus and *Lactobacillus jensenii* 5L08 showed antagonistic activity against *G. vaginalis* through competition and bacteriocin production [34].

Future application for lantibiotics may include the development of new peptides based on desirable structural features of some well-characterized biologically active peptides such as nisin. The observation that nisin binds to an invariable region of lipid II indicates that this bacteriocin could be used as a template for designing new drugs, since this is the same target for vancomycin resistance. The absence of post-translational modifications in class II bacteriocins makes them more amenable to change, a fact that is best illustrated by the creation of hybrid bacteriocins. The natural association between bacteriocin biosynthesis genes and plasmids can be exploited to facilitate heterologous bacteriocin production [1]. Bioengineered peptides derived from bacteriocin-producing strains have already been produced heterologously in *E. coli* and yeasts [35, 36].

Selecting narrow-spectrum antimicrobial can help address the current threat posed by multi-resistance bacterial pathogens. It leads to a reduction in the collateral killing of nonpathogenic species, that is, commensal species, which in turn leads to a decrease in nosocomial infection levels [10].

The potential for development of resistance against bacteriocins is a cause of concern if they are to be used as food preservatives. Most research in this field has focused on nisin. Gravesen *et al.* (2002) [37] studied the frequency of spontaneous resistance to nisin in *L. monocytogenes* varies from 10^{-2} to 10^{-7}. The specific mechanism(s) by which cells become resistant to nisin is not well understood, although it is apparently not related to variations in the lipid II content. The well-studied mechanism of resistance to class II bacteriocins is in *L. monocytogenes*, in which resistance seems to be linked to reduced expression of a mannose permease of the phosphotransferase system. Similar to nisin-resistant mutants, other factors such as cell membrane fluidity and cell surface charge impact on resistance. Further studies are required to determine the frequency of resistance to other bacteriocins and the risk of its development in large-scale industrial environment. Resistance might best be limited by controlling several factors, such as pH variations, osmolarity and temperature [1].

6.3. In Livestock Health

Antibiotics have been used as therapeutic and prophylactic treatments to control a variety of bacterial infections in livestock for several decades. Different types of antibiotics have been fed at sub-therapeutic levels to cattle, poultry and swine to increase productivity and feed efficiency. The emergence of antibiotic-resistance in many bacteria relevant for animal and public health stresses the importance of decreasing the use of antibiotics in animal production. The reduction of antibiotic application in livestock can only be achieved if alternative antimicrobial strategies are available. Those alternative strategies can be immunization, diet modification, sanitation, feed additives, and probiotic bacteria. A variety of probiotic bacteria have been tested to control animal and foodborne pathogenic bacteria in livestock, but in many of them the beneficial effects have not been fully elucidated. The mechanism of inhibition of pathogenic bacteria for several of those probiotic microorganisms is mediated by the production of bacteriocins. The application of bacteriocins such as colicins in livestock has been largely achieved by feeding bacteriocin producing strains. The use of bacteriocins in livestock is largely based on feeding or applying bacteriocin-producing bacteria (BPB) to the diet [38]. To maintain intestinal microbiota balance in agricultural animals, it is important to control the overgrowth of potentially pathogenic bacteria. Probiotic supplements are a safe alternative to traditional antibiotics or growth factors. There are several bacteriocin producing cultures available commercially. Such examples include Bovamine®, a mixture of *Lactobacillus* strains has shown promising results to reduce the fecal shedding of *E. coli* O157:H7 in cattle and is currently being marketed as a probiotic. Several strains of colicinogenic *E. coli* (CEC) have also been commercialized. The utilization of bacteriocins or bacteriocin-producing bacteria in livestock is a field with enormous possibilities for both research and commercialization.

7. "Raison d'etre" - Ecological Role and Evolution of Bacteriocin Diversity

There is a lot of discussion regarding the role of bacteriocins in nature. It seems that the prime reason is to provide an ecological advantage to the producing organisms over its competitor [9]. Most bacteriocins are structurally complex and therefore come at a considerable cost for the producer cell, they must be functionally indispensable for the existence and persistence in nature of bacterial and archaeal cell lineages. However, exposure of producing bacteria to laboratory conditions (no competitors and low stress) can lead to the elimination of bacteriocin-encoding plasmids. There must be a tradeoff between the metabolic (and genetic) cost to bacteriocin production and the survival benefits derived from both the expression and/or retention of the genetic capability of producing bacteriocin(s) as the need arises [2].

The dissemination of bacteriocin production among LAB is undoubtedly related to the fact that these genes are often associated with transferable elements such as conjugative transposons or plasmids [1].

Observations on the dissemination of bacteriocins, points to them having a critical role in mediating microbial population or community interactions. Nonetheless, exactly what is their role? Bacteriocins may serve as anticompetitors, enabling the invasion of a strain into an established microbial community. They may also play a defensive role and act to prohibit the invasion of other strains or species into an occupied niche or limit the advance of neighboring cells. It has recently been proposed that Gram-positive bacteriocins mediate quorum sensing mechanisms [10, 11].

CONCLUDING REMARKS

For the past 60 years, bacteriocins have frequently been proposed as the solution for food contamination problems. However, nisin and, to less extent, pediocin PA1, remain the only examples of the large-scale deliberate use of these peptides in food applications.

Although intensive studies over the last decade have greatly advanced our knowledge base about bacteriocins, further work is needed before we are able to fully understand the molecular mechanisms, structure-function relationships, and mechanisms of action of bacteriocins.

Genomic advances will also contribute to the identification of new bacteriocins. A better understanding of regulatory mechanisms will allow controlling production when desired and even how to generate overproducing strains. By determining the specific activity of bacteriocins against particular spoilage/pathogenic bacteria, ensuring stability and assessing the likelihood of emerging resistance, it should be possible to pre-determine which bacteriocin is best suited for a particular food application.

Bacteriocins represent one of the most common microbial defense systems. Although we are still in the earliest stages of exploring their evolutionary relationships and ecological roles, it is clear from their abundance and diversity that they are the microbial weapons of choice. Sorting out why they are such a successful family of toxins will require a substantial commitment to future research. In addition, we require more sophisticated ecological models (both empirical and theoretical) to aid in our growing sense of the diverse roles the toxins play in mediating microbial dynamics and maintaining microbial diversity. The impact of such studies is not solely academic. The potential for bacteriocins to serve as alternatives to classical antibiotics in treating bacterial infections is real, and the application of bacteriocins in food preservation is on the rise. The future role bacteriocins may serve are only limited by our own imagination.

CONFLICT OF INTEREST

The authors confirm that this chapter contents have no conflict of interest.

ACKNOWLEDGEMENTS

Declared none.

REFERENCES

[1] Cotter PD, Hill C, Ross RP. Bacteriocins: developing innate immunity for food. Nat Rev Microbiol 2005; 3(10): 777-88.
[http://dx.doi.org/10.1038/nrmicro1273] [PMID: 16205711]

[2] Heng NCK, Wescombe PA, Burton JP, Jack RW, Tagg JR. The diversity of bacteriocins in Gram-positive bacteria In: MA Riley, MA Chavan, Eds. Bacteriocins: Ecol Evol. 2007; pp. 45-92.

[3] Nes IF, Diep DB, Ike Y. Enterococcal bacteriocins and antimicrobial proteins that contribute to niche control. In: Gilmore MS, Clewell DB, Ike Y, Shankar N, Eds. Enterococci: From commensals to leading causes of drug resistant infection. Boston: Massachusetts Eye and Ear Infirmary 2014; pp. 1-15.

[4] Tagg JR, Dajani AS, Wannamaker LW. Bacteriocins of gram-positive bacteria. Bacteriol Rev 1976; 40(3): 722-56.

[PMID: 791239]

[5] Klaenhammer TR. Genetics of bacteriocins produced by lactic acid bacteria. FEMS Microbiol Rev 1993; 12(1-3): 39-85.
[http://dx.doi.org/10.1111/j.1574-6976.1993.tb00012.x] [PMID: 8398217]

[6] Gratia A. Sur un remarquable exemple d'antagonisme entre deux souches de colibacille. C. R. Soc Biol 1925; 93: 1040-2.

[7] Fredericq P. Sur la pluralité des récepteurs d'antibiose de *E. coli.* C. R. Soc Biol 1946; 140: 1189-90.

[8] Klaenhammer TR. Bacteriocins of lactic acid bacteria. Biochimie 1988; 70(3): 337-49.
[http://dx.doi.org/10.1016/0300-9084(88)90206-4] [PMID: 3139051]

[9] Riley MA, Wertz JE. Bacteriocins: evolution, ecology, and application. Annu Rev Microbiol 2002; 56: 117-37.
[http://dx.doi.org/10.1146/annurev.micro.56.012302.161024] [PMID: 12142491]

[10] Riley M. Bacteriocins, biology, ecology, and evolution. *Encyclopedia of Microbiology.* Available at: http://works.bepress.com/margaret_riley/6/ , 2009 [Accessed December 29, 2014];
[http://dx.doi.org/10.1016/B978-012373944-5.00065-1]

[11] Riley M, Chavan M. Bacteriocins: ecology and evolution. In: M.A. Riley, M.A. Chavan, Eds. Berlin, Heidelberg: Springer 2006.

[12] Jack RW, Jung G. Lantibiotics and microcins: polypeptides with unusual chemical diversity. Curr Opin Chem Biol 2000; 4(3): 310-7.
[http://dx.doi.org/10.1016/S1367-5931(00)00094-6] [PMID: 10826980]

[13] Duquesne S, Petit V, Peduzzi J, Rebuffat S. Structural and functional diversity of microcins, gene-encoded antibacterial peptides from enterobacteria. J Mol Microbiol Biotechnol 2007; 13(4): 200-9.
[http://dx.doi.org/10.1159/000104748] [PMID: 17827970]

[14] Kleanthous C. Swimming against the tide: progress and challenges in our understanding of colicin translocation. Nat Rev Microbiol 2010; 8(12): 843-8.
[http://dx.doi.org/10.1038/nrmicro2454] [PMID: 21060316]

[15] Lindgren SE, Dobrogosz WJ. Antagonistic activities of lactic acid bacteria in food and feed fermentations. FEMS Microbiol Rev 1990; 7(1-2): 149-63.
[http://dx.doi.org/10.1111/j.1574-6968.1990.tb04885.x] [PMID: 2125429]

[16] Chen H, Hoover DG. Bacteriocins and their food applications. Comp Rev Food Sci Food Safe 2003; 2(3): 82-100.
[http://dx.doi.org/10.1111/j.1541-4337.2003.tb00016.x]

[17] Jacob F, Lwoff A, Siminovitch A, Wollman E. Définition de quelques termes relatifs a la lysogénie. Ann Inst Pasteur (Paris) 1953; 84(1): 222-4.
[PMID: 13031254]

[18] Jack RW, Tagg JR, Ray B. Bacteriocins of gram-positive bacteria. Microbiol Rev 1995; 59(2): 171-200.
[PMID: 7603408]

[19] Kemperman R, Jonker M, Nauta A, Kuipers OP, Kok J. Functional analysis of the gene cluster

involved in production of the bacteriocin circularin A by Clostridium beijerinckii ATCC 25752. Appl Environ Microbiol 2003; 69(10): 5839-48.
[http://dx.doi.org/10.1128/AEM.69.10.5839-5848.2003] [PMID: 14532033]

[20] Hansen JN. Nisin as a model food preservative. Crit Rev Food Sci Nutr 1994; 34(1): 69-93.
[http://dx.doi.org/10.1080/10408399409527650] [PMID: 8142045]

[21] Kok J, Holo H, van Belkum MJ, Haandrikman AJ, Nes IF. Bacteriocins of Lactic Acid Bacteria. Elsevier 1993.

[22] Galvin M, Hill C, Ross RP. Lacticin 3147 displays activity in buffer against gram-positive bacterial pathogens which appear insensitive in standard plate assays. Lett Appl Microbiol 1999; 28(5): 355-8.
[http://dx.doi.org/10.1046/j.1365-2672.1999.00550.x] [PMID: 10347889]

[23] Kuipers OP, Beerthuyzen MM, de Ruyter PG, Luesink EJ, de Vos WM. Autoregulation of nisin biosynthesis in Lactococcus lactis by signal transduction. J Biol Chem 1995; 270(45): 27299-304.
[http://dx.doi.org/10.1074/jbc.270.45.27299] [PMID: 7592991]

[24] Nes IF, Diep DB, Håvarstein LS, Brurberg MB, Eijsink V, Holo H. Biosynthesis of bacteriocins in lactic acid bacteria. Antonie van Leeuwenhoek 1996; 70(2-4): 113-28.
[http://dx.doi.org/10.1007/BF00395929] [PMID: 8879403]

[25] Franz CM, van Belkum MJ, Holzapfel WH, Abriouel H, Gálvez A. Diversity of enterococcal bacteriocins and their grouping in a new classification scheme. FEMS Microbiol Rev 2007; 31(3): 293-310.
[http://dx.doi.org/10.1111/j.1574-6976.2007.00064.x] [PMID: 17298586]

[26] Kleerebezem M. Quorum sensing control of lantibiotic production; nisin and subtilin autoregulate their own biosynthesis. Peptides 2004; 25(9): 1405-14.
[http://dx.doi.org/10.1016/j.peptides.2003.10.021] [PMID: 15374644]

[27] Hoover DG, Harlander SK. Bacteriocins of Lactic Acid Bacteria. Elsevier 1993.

[28] Ray B, Hoover DG. Bacteriocins of Lactic Acid Bacteria. Elsevier 1993.

[29] Appendini P, Hotchkiss JH. Review of antimicrobial food packaging. Innov Food Sci Emerg Technol 2002; 3(2): 113-26.
[http://dx.doi.org/10.1016/S1466-8564(02)00012-7]

[30] Giraffa G. Enterococci from foods. FEMS Microbiol Rev 2002; 26(2): 163-71.
[http://dx.doi.org/10.1111/j.1574-6976.2002.tb00608.x] [PMID: 12069881]

[31] Mota-Meira M, LaPointe G, Lacroix C, Lavoie MC. MICs of mutacin B-Ny266, nisin A, vancomycin, and oxacillin against bacterial pathogens. Antimicrob Agents Chemother 2000; 44(1): 24-9.
[http://dx.doi.org/10.1128/AAC.44.1.24-29.2000] [PMID: 10602718]

[32] Reddy KV, Aranha C, Gupta SM, Yedery RD. Evaluation of antimicrobial peptide nisin as a safe vaginal contraceptive agent in rabbits: *in vitro* and *in vivo* studies. Reproduction 2004; 128(1): 117-26.
[http://dx.doi.org/10.1530/rep.1.00028] [PMID: 15232069]

[33] Di Pierro F, Colombo M, Zanvit A, Risso P, Rottoli AS. Use of *Streptococcus salivarius* K12 in the prevention of streptococcal and viral pharyngotonsillitis in children. Drug Healthc Patient Saf 2014; 6: 15-20.
[http://dx.doi.org/10.2147/DHPS.S59665] [PMID: 24600248]

[34] Kaewsrichan J, Chandarajoti K, Kaewnopparat K, Kaewnopparat N. Evaluation of *Lactobacilli* containing suppository formulation for probiotic use. Mahidol Univ J Pharm Sci 2007; 34(1-4): 1-8.

[35] Schoeman H, Vivier MA, Du Toit M, Dicks LM, Pretorius IS. The development of bactericidal yeast strains by expressing the *Pediococcus acidilactici* pediocin gene (pedA) in *Saccharomyces cerevisiae*. Yeast 1999; 15(8): 647-56.
[http://dx.doi.org/10.1002/(SICI)1097-0061(19990615)15:8<647::AID-YEA409>3.0.CO;2-5] [PMID: 10392443]

[36] Richard C, Drider D, Elmorjani K, Marion D, Prévost H. Heterologous expression and purification of active divercin V41, a class IIa bacteriocin encoded by a synthetic gene in *Escherichia coli*. J Bacteriol 2004; 186(13): 4276-84.
[http://dx.doi.org/10.1128/JB.186.13.4276-4284.2004] [PMID: 15205430]

[37] Gravesen A, Jydegaard Axelsen A-M, Mendes da Silva J, Hansen TB, Knøchel S. Frequency of bacteriocin resistance development and associated fitness costs in Listeria monocytogenes. Appl Environ Microbiol 2002; 68(2): 756-64.
[http://dx.doi.org/10.1128/AEM.68.2.756-764.2002] [PMID: 11823216]

[38] Diez-Gonzalez F. Applications of bacteriocins in livestock. Curr Issues Intest Microbiol 2007; 8(1): 15-23.
[PMID: 17489435]

CHAPTER 8

Biocides – A Reasonable Alternative to Prevent and Control Microorganisms?

Rui Seixas[*], **Carla Mottola**

CIISA/Faculty of Veterinary Medicine of the University of Lisbon, Avenida da Universidade Técnica, 1300-477 Lisbon, Portugal

Abstract: Acute infections caused by pathogenic microorganisms have been studied extensively for over 100 years. Infections that killed millions of people in the past are now ever more successfully controlled due to the development and use of antimicrobial compounds. Biocides, in particular, have been used in various forms for many centuries to control the dissemination of microorganisms and represent a wider group of substances with a key role in the medical and food industry. This chapter presents the main biocidal classes and mechanisms of action, including their respective spectrums of activity and use in several fields. Factors influencing the effectiveness of biocides are addressed as well as biocide resistance and the link with antibiotic resistance. Several mechanisms of biocide resistance are presented on a molecular level with a particular focus on biofilm resistance and its formation in bacterial populations. This chapter concludes with an introduction to ECHA, the European Chemicals Agency and their recently approved regulatory framework for biocide products in Europe.

Keywords: Antimicrobial Resistance, Biocides, Biofilm, Concentration, Contact Time, Cross-Resistance, Disinfection, ECHA, Effectiveness, European Union, Exopolymeric products, Legislation, Microorganisms, pH, Quaternary ammonium compounds, REACH, Regulation, *Salmonella*, *Staphylococcus aureus*, Temperature.

1. INTRODUCTION

Antimicrobial resistance in bacterial pathogens and other microorganisms is disseminated worldwide leading to therapeutical failures in both human and animal diseases. This situation poses an emergent challenge in research, aiming at

[*] **Corresponding author Rui Seixas:** CIISA/Faculty of Veterinary Medicine of the University of Lisbon, Avenida da Universidade Técnica, 1300-477 Lisbon, Portugal; Tel. + 351 213852800; Email: adruiseixas@fmv.ulisboa.pt

Manuela Oliveira and Isa Serrano (Eds)

the identification of new approaches to control multidrug resistant pathogens by discovering new compounds or increasing the effectiveness of pre-existing ones such as antibiotics or biocides [1, 2].

The biocides have been used in various forms for centuries. These compounds are chemical substances with the ability to impair, render harmless, or apply a controlling effect on any microorganism by chemical or biological means. Biocides represent a wider group of substances with a key role in pharmaceutical and medical industry [3]. They are also commonly used in many other fields like agriculture or even forestry [4 - 6].

In the past, several references were made regarding the use of these chemical substances: empirical approaches of copper and silver vessels for storing potable water, vinegar and honey for disinfecting wounds or even balsams as preservatives in mummification. Later, reports were made about the use of iodine as a wound disinfectant, of chlorine water in obstetrics and the claimed sporicidal activity of mercuric chloride [7, 8].

In the beginning of the twentieth century, biocides such as chlorine agents and some quaternary ammonium compounds (QACs) were presented [9]. In 1945, several groups of biocides were commonly used, in particular: phenolics, organomercurials, chlorine agents, iodine, alcohols, formaldehyde, hydrogen peroxide and silver compounds [8].

Nowadays, probably one of the most important agents introduced, especially in the medical industry, are the biguanides, namely chlorhexidine. This compound is a cationic polybiguanide with antifungal and bactericidal properties [10]. Its importance was soon recognized in 1979, when chlorhexidine was considered by the World Health Organization in the List of Essential Medicines as a critical compound needed in a basic health system [5, 11].

Today, the interest and need for biocides is not diminished. On the contrary, its use has expanded and it is completely widespread in all types of industries. In 2006, the global biocide demand was estimated at €10-11 billion with a growth of 4-5% per annum for the previous 15 years, and the market expansion is predicted to continue. Despite the increased demand in biocides, application and selection of biocides has become more restricted in the last three decades. These restrictions were applied in response of two main concerns: product toxicity and environmental impact [12].

Regardless market constraints, biocides are promising compounds being effective against virulent and antimicrobial resistant bacteria such as *Escherichia coli* O157:H7, methicillin-resistant *Staphylococcus aureus* (MRSA) or multidrug resistant *Salmonella* spp. [13 - 15]. These substances can also be active against fungi [16]. Biocides also show an efficacy against bacterial biofilms, one of the major reasons for chronic relapsing infections, impairing biofilm formation [17]. These compounds show a great ability to deal with resistant microorganisms, and the emerging resistance against different types of biocides has only been studied and characterized recently [18, 19].

New antimicrobial therapeutic options to treat multi-resistant bacteria are progressively decreasing and biocides may be a promising alternative to impair their dissemination [20, 21]. The purpose of this chapter is to review biocides classes, mechanism of action and effectiveness with a particular focus on the mechanisms of biocide resistance, including biofilm.

2. BIOCIDES CLASSES

"Biocide" describes a chemical agent, natural or synthetic, that inactivates microorganisms and usually has a broad spectrum [22]. The antimicrobial activity of these compounds can range from inhibiting the target growth, in this case we use the term "-static," to kill the target, the term "-cidal". Biocides include antiseptics, disinfectants and preservatives. Antiseptics are products largely used in the medical field that inhibit the growth or the development of microorganisms in living tissue, not necessarily killing, such as oral hygiene and wounds cleaning. Disinfectants are generally used on inanimate objects or surfaces because they can injure the tissues, inactivate most microorganisms but not commonly spores, and are used in many applications in daily life.

Preservatives act to prevent the microbe's multiplication in formulated products, including foods and pharma-ceuticals [22 - 24].

In order to perform a biocidal classification, it is important to elucidate the term "sterilization", which is defined in this chapter, as a chemical or physical process that destroy all forms of life including spores [25, 26].

The mode of action of these active agents is still not fully clear, but it is known that biocides have a broader spectrum activity compared to chemotherapeutic agents and it is generally accepted that they have multiple targets in microorganisms, while antibiotics have specific intracellular targets. For this reason, the overall damage of biocides results in a bactericidal effect [22, 23]. Generally, lower concentrations of biocide result in a bacteriostatic effect, corresponding to a reversible activity on the impairment of enzymatic activity and/or on the cytoplasmic membrane, and whose mechanism is poorly documented, anticipating that a unique target site within the cell is involved [22]. Some studies have been confirming this hypothesis, for example triclosan has been found to block lipid synthesis by specifically targeting the enoyl-acyl carrier protein reductase in *E.coli* and *Mycobacterium smegmatis* [27].

The efficacy of the biocides depend on the type of microorganism [25]. Nowadays, a scale provided by Spaulding [28] in 1957 and revised by Favero and Bond [29] in 1991, which orders the microorganisms based on their sensitivity to biocides, and consequently the required levels of disinfection, is still in use. In 2011, McDonnell and Burke raised the issue that despite Spaulding classification is still valid, possibly the expectations of the disinfection levels need to be additionally tested, taking into account the new available methods and the advances on microbiological studies [25].

Biocides mechanism reflects their diversity in chemical structure and their physiology. The first contact with the microbial cell is binding to the cell surface. Subsequently, the biocide adsorption leads to cell penetration reaching its target site by one or more interaction that can be done with outer cell components, or with the cytoplasmic membrane or its constituents [23, 27, 30].

Biocides include a variety of classes (Table **1**) that differ in chemical structure, mechanism of action, toxicity, efficacy, risks and use in different situations [22, 26]. Many of them are no longer used today, because of their allergic properties, like as mercury compounds [24]. Furthermore a brief resume of the most important classes of biocides are introduced:

2.1. Alcohols

Largely used for skin antisepsis and surface disinfection, but not for sterilization because they are reversible inhibitors of sporulation. They are active against viruses, bacteria including mycobacteria, and fungi. Examples of alcohols are ethyl and n-propanol, which are the most used. Their mechanism of action is still unclear, but it seems that they cause damage to cytoplasmic membrane interfering with its permeability, electron transport chain, and denature proteins [27, 31].

2.2. Aldehydes

Aldehydes include glutaraldehyde, a dialdehyde that has a large activity spectrum on bacteria and their spores, viruses and fungi. This biocide is mainly used in sterilization of surgical material and endoscopes, and for low-temperature disinfection [22, 26]. It is more powerful at high temperatures and pH greater than 7 [26]. Its mechanism of action consists in binding to the outer layers of microorganisms, inhibition of periplasmic enzymes, DNA, RNA, and protein synthesis [22, 27, 32]. Aldehydes also include formaldehyde, a monoaldehyde with the same spectrum and mechanism of action of glutaraldehyde but acting more slowly. It is used generally in disinfection and sterilization in liquid or in combination with low temperature steam (since it exists as a freely water-soluble gas), as like as in decontamination of electronic and mechanic equipment. *o*-Phthalaldehyde is a new type of disinfectant that seems to have potent bactericidal and sporicidal activity, and it may replace glutaraldehyde in endoscope disinfection [22, 26]. Despite their efficacy and large spectrum, they have the disadvantage to be highly irritant and potentially carcinogenic [33].

2.3. Anilides

Anilides are rarely used in the clinic and the most studied compound is triclocarban, especially active against Gram-positive bacteria and much less against Gram-negative bacteria and fungi. It is used especially in personal care products and consumer soaps [34].

Anilides act on the cytoplasmic membrane, modifying its permeability and leading to cellular death [22, 27]. Triclocarban can persist in the environment and exhibit toxicity effects, like accumulating on aquatic systems [35], and it has showed adverse health effects in animal models [34].

2.4. Biguanides

This class includes chlorhexidine, a bactericidal and non-sporicidal agent with broad-spectrum efficacy in particular for the skin, causing low irritation. It is greatly reduced in the presence of organic matter and is pH-dependent [26, 36]. Probably it is the most used biocide in antiseptic products, mainly in handwashing and oral products, but also as a disinfectant and preservative. Its lethal effects are due to alterations of the membrane permeability [22, 32]. Chlorhexidine is only effective on viruses that are lipid-enveloped. Alexidine differs chemically from chlorhexidine in possessing ethylhexyl end groups and it also works faster. Polymeric biguanides, like vantosil, are largely used for disinfection of swimming pools and as general disinfectants agents in the food industry [37]. Biguanides are toxic to fish and should not be discharged into the environment [26].

2.5. Diamidines

Diamidines like propamidine, pentamidine, dibromopropamidine isethionates is a class of biocides characterized by low molecular weight and synthetic producibility, so they are very appealing [38]. Their use in protozoan infections dates back a long time and there are many reports about their activity in some species of *Plasmodium*, *Leishmania*, and more recently *Babesia*. [38, 39]. Clinically, diamidines are used for the topical treatment of wounds. Their mechanism is already unclear but they seem to inhibit oxygen uptake and induce leakage of amino acids. It has been already shown that the aromatic diamidine DB75 can collapse the mitochondrial membrane potential [40].

2.6. Halogen-Releasing Agents

These compounds are used in clinic antiseption and disinfection. The most used halogen compounds are chlorine-releasing agents (CRA) and iodine-releasing agents (IRA). They lose activity rapidly in the presence of sunlight, organic matter and some metals, so they must be applied to cleaned surfaces for disinfection [26].

The most important CRAs include sodium hypochlorite, chlorine dioxide, and the N-chloro compounds such as sodium dichloroisocyanurate [22]. These oxidizing compounds are mainly used for hard-surface disinfection, including blood containing human immunodeficiency virus or hepatitis B virus. They act destroying the cellular activity of proteins and have a sporicidal activity even in low concentrations [41].

Iodine is rapidly bactericidal, fungicidal, tuberculocidal, virucidal and sporicidal, but iodine solutions cause irritation and are unstable. For these reasons, researchers have developed IRA, also called iodophors that act as reservoirs of the active free iodine. The most used as both antiseptics and disinfectants are povidone-iodine and poloxamer-iodine. Iodine rapidly penetrates into microorganisms and attacks key groups of proteins, nucleotides and fatty acids, which lead to cell death. Enveloped viruses are also susceptible to surface proteins attacks of iodine [22].

2.7. Silver Compounds

Silver compounds such as silver metal, silver acetate and silver nitrate have an important antimicrobial activity. Nowadays silver sulfadiazine (AgSD) is the most important silver compound in use. In the last years, silver compounds have been used as antimicrobials in burns, warts, diabetic ulcers and trauma [42]. AgSD is a combination of two antibacterial agents, silver and sulfadiazine; it has a broad spectrum activity and binds to cell components, including DNA [22, 43].

2.8. Peroxygens

Peroxygens include the hydrogen peroxide (H_2O_2) that has a broad-spectrum efficacy against viruses, bacteria (especially Gram-positive), yeasts and bacterial spores. It is frequently used for disinfection, sterilization and antisepsis. Sporicidal activity requires higher concentrations and longer contacts times. H_2O_2 attacks essential cell components, including lipids, proteins and DNA, acting as an oxidant by producing hydroxyl free radicals. It also interact with ribosomes, although they are not the primary site [27, 32].

Another peroxygen biocide more potent than hydrogen peroxide is the peracetic acid (PAA) that is bactericidal, sporicidal, virucidal, and fungicidal at low concentrations. Like H_2O_2, it probably denatures proteins and enzymes and increases cell wall permeability [26, 36]. Its applications include sterilization of medical devices and sterilization of the environmental surface [22].

2.9. Phenols

Antimicrobial agents used for their antiseptic and disinfectant properties. They are fungicidal, bactericidal, tuberculocidal and virucidal for enveloped viruses, and not sporicidal [26, 44]. They induce a progressive leakage of intracellular constituents, inclusive denaturation of proteins, leading to membrane damage [22]. Phenols are generally safe for humans, but prolonged use can be irritant for skin, and they are highly toxic for animals, especially for cats [26].

2.10. Bis-Phenols

Bis-Phenols are hydroxy-halogenated derivatives of two phenolic groups connected by various bridges. They have a broad-spectrum efficacy but little activity against *Pseudomonas aeruginosa* and are sporostatic. The most used are triclosan and hexachlorophene that have cumulative and lasting effects on the skin. Triclosan has a particular activity against Gram-positive bacteria and also has anti-inflammatory activity; at low concentrations, it inhibits the uptake of essential nutrients, whereas at higher concentrations it is bactericidal, causing rapid release of cellular components and cell death [22]. It is widely used mainly in toothpaste, textile and sport equipment. Triclosan can persists in the environment exhibiting toxicity effects [34, 35]. Hexachlorophene is another bis-phenol with a broad spectrum, but due to its toxicity, particularly in neonates, its use has been limited [22].

2.11. Halophenols

Chloroxylenol is an example of a halophenol, frequently used in antiseptic and disinfectant formulations. It is bactericidal but *P. aeruginosa* is highly resistant to it. Because of its phenolic nature, it would be expected to has an effect on microbial membranes [22].

2.12. Quaternary Ammonium Compounds

Quaternary Ammonium Compounds are cationic agents, surface-active (surfactants), with two regions in their molecular structures, a hydrocarbon water-repellent and a water-attracting hydrophilic group. They irreversibly bind to phospholipids in the cell membrane and denaturate proteins, impairing the cell permeability [26, 44]. In addition to antimicrobial activity, they are also excellent for hard-surface cleaning and deodorization, and they are not irritant to skin. QACs are highly active against Gram-positive bacteria and have a good efficacy against Gram-negative bacteria, fungi and enveloped viruses, including human immunodeficiency virus and hepatitis B virus (not in non-enveloped viruses) and are sporostatic [26, 36].

2.13. Vapor-Phase Sterilants

Biocides such as vapor-phase sterilants can guarantee the sterilization of many heat-sensitive medical devices and surgical supplies by using liquid or vapor-phase sterilization systems. The most widely used agents in these systems are ethylene oxide, formaldehyde and alkylating agents both with broad spectrum, however their activity depend on some factors like temperature and exposure time [22]. The ethylene oxide sterilization method increased exponentially and today is the most important sterilization method for medical devices. Its alkylation reaction with cellular constituents of organisms results in proteins and enzymes denaturation and DNA damage [45, 46].

Table 1. Summary of biocides classes and their action mechanism.

Class	Antiseptic or Disinfectant	Target
Alcohols	Ethanol Isopropanol	Cytoplasmic membrane (increased permeability)
Aldehydes	Glutaraldehyde Formaldehyde	Cross linking of macromolecules
Anilides	Triclocarban	Cytoplasmic membrane (membrane potential and electron transport chain)
Biguanides	Chlorhexidine Alexidine	Cytoplasmic (inner) membrane

(Table 1) contd.....

Class	Antiseptic or Disinfectant	Target
Bisphenols	Triclosan Hexachlorophene	Cytoplasmic membrane (increased permeability)
Diamidines	Propamidine Dibromopropamidine	Cytoplasmic (inner) membrane
Halogen-releasing agents	Chlorine compounds Iodine compounds	Effects on DNA
Halophenols	Chloroxylenol (PCMX)	Cytoplasmic (inner) membrane
Heavy metal derivatives	Silver compounds	Interactions with thiols groups
Peroxygens	Hydrogen peroxide Ozone Peracetic acid	Effects on DNA
Phenols and cresols	Phenol Cresol	Cytoplasmic (inner) membrane
Quaternary ammonium compounds	Cetrimide Benzalkonium chloride	Cytoplasmic (inner) membrane
Vapor phase	Ethylene oxide Formaldehyde Hydrogen peroxide	Cytoplasmic membrane Cross linking of macromolecules Effects on DNA

More recently the hydrogen peroxide and paracetic acid are also used in these vapor-phase systems and considered more active, as oxidants, at lower concentrations than in the liquid form. Their main advantages over other vapor-phase systems include low toxicity, rapid action and activity at lower temperature; the disadvantages are limited penetrability and applicability since paracetic acid can cause corrosion to certain materials and irritation with prolonged use [47].

3. FACTORS INFLUENCING EFFECTIVENESS OF BIOCIDES

Biocides effectiveness is not only dependent on their action mechanism but also on their application conditions [48]. The efficacy of biocides greatly depends on several factors such as contact time, concentration, temperature, pH and microorganisms type [23, 49]. All biocides types show a range of activity against bacteria, bacterial spores, fungi, viruses and protozoa and some have algicidal activity [23, 50]. Biocides efficacy depends on intrinsic and extrinsic factors. The intrinsic factors are characterized by the biocide type and its concentration, contact time and application [44, 49]. For example, the efficacy of disinfectants against biofilms or planktonic bacteria usually increases with the increase in both biocide concentration and the contact time of treatment [51 - 53]. However, these factors require a balance between efficacy and toxicity for human or animal cells. This balance is also important in other types of industry, for example, some disinfectants in high concentrations are corrosive to metals, including stainless steel. However, when diluted to their concentration use (1000 ppm or less), most are non-corrosive. The exceptions are oxidizing biocides, which are strongly corrosive even at use levels (1 ppm or less) [54].

Biocides concentration and contact time influence a microbial population and the relationship between these two variables determines the microbial reduction. The stability of the active biocide compounds is also importance for the biocide action upon the microbial population. The temperature and pH are crucial, as most substances have a higher efficiency at high temperature and pH [51]. The contact mode also influences the biocides efficacy, as well as the contact time which plays an important role. Extrinsic factors consist in the presence of organic matters, the surface type, the microorganisms incubation time and concentration and the microorganism's ability to transform the biocide into inactive forms [22]. Organic matter is known to reduce the activity of the biocides [55] and the surface type may influence the efficacy of biocides in killing and removing biofilms from the abiotic surfaces [56]. Incubation time also plays a role in biocide efficacy since the increase in the incubation time of biofilms usually results in an increase of biofilm resistance to biocides [57].

4. MECHANISMS OF BIOCIDE RESISTANCE

The use of antiseptic and disinfectant formulations has increased over the years, leading to the development of microbial resistance and cross-resistance to antibiotics.

Biocide resistance is an old phenomenon first recognized seventy years ago by the identification of chlorine resistance in *Salmonella Typhi* [58]. However, the link with the penicillin resistance was only recently identified [58]. Is still a current discussion if the biocide use may have impact on the prevalence of antibiotic resistant microorganisms, yet epidemiological data is not sufficient to explain this association. One example of biocide resistance linked to multiple antibiotic resistance is the *mar* protein: strains that express this protein are resistant to tetracycline, chloramphenicol, triclosan and also to pine oil [59].

A strain is defined resistant to a biocide if it is not inactivated by a specific concentration or time of exposure that usually inactivates the majority of other strains [60]. It is important to note that, unlike antibiotics, the resistance to a biocide does not necessarily correlates to a therapeutic failure [22].

Resistance can be a natural property of an organism (intrinsic) or acquired (extrinsic) by mutation or through the acquisition of plasmids or transposons. Intrinsic resistance is showed by Gram-negative bacteria, bacterial spores, mycobacteria, and under certain conditions, in staphylococci [22, 59]. Acquired plasmid mediated resistance is mostly associated with mercury compounds and other metallic salts, and it has been observed in staphylococci [22].

4.1. Intrinsic Resistance Mechanisms

The microorganism outer membrane is the first challenge for a disinfectant or antiseptic. The bacterial spores have a complex structure in which the germ cell and germ cell wall are surrounded by a cortex constituted by the inner and outer spore coats and this is an important aspect in the resistance to biocides [50]. Regarding Mycobacteria, their high resistance is partly due to the complex cell wall, a highly hydrophobic structure with a mycoylarabinogalactan-peptidoglycan skeleton that provides an effective barrier to the entry of these agents [22, 61]. The cell wall of staphylococci does not seem a really barrier to the biocide action, whereas the biofilm, which will be discussed later, may play a protective role [62]. Enterococci are generally less sensitive to biocides than staphylococci [22]. On the other hand, the Gram-negative bacteria have an outer membrane limiting the penetration of many antibacterial products because of its hydrophobic molecules. *P. aeruginosa* is considered one of the highest resistant Gram-negative bacteria, due to strong links between lipopolysaccharides and the small size of porins that impair the biocide diffusion [22, 63].

4.2. Biofilm As an Intrinsic Resistance Mechanism

One of the most important intrinsic mechanisms consisting in a physiological adaptation to the biocide action, with the ability to confer an increased resistance to biocides, is the biofilm formation. Today, the success of most bacterial infections relies on the bacterial ability to establish adherent bacterial communities on surfaces [64]. These bacterial communities are called biofilms. They can be defined as a community of microorganisms which are irreversibly attached to biotic or abiotic surfaces enclosed in an extracellular matrix composed of exopolymeric products [65]. This matrix, mainly produced by the micro-organisms themselves, is a conglomeration of extracellular polymeric substances such as nucleic acids, proteins polysaccharides and others [66].

There are many reasons for the reduced sensitivity of the bacteria in a biofilm, such as the production of degradative enzymes, genetic exchanges between microorganisms within the biofilm and reduced penetration of the biocide in the cells. These factors create a new microbiological environment favorable to resistance. Nowadays, one of the biggest concerns in microbiology is the understanding of the mechanisms involved in biofilm resistance to biocides. The Minimum Biofilm Eradication Concentration assay has recently been approved by the American Society for Testing and Materials as a standard method to evaluate the effectiveness of antimicrobials in biofilms [3, 67]. Moreover, in most cases, biofilms are complex mixtures of different species that are generally more resistant to disinfection than one single species biofilms. It was proposed that in polymicrobial biofilms, the chemical interactions between the polymers may lead to a more viscous matrix and the production of the enzymes by different species act synergistically against toxic compounds [68]. In the biofilm spatial structure, some strains may be protected from the antimicrobial compounds [3].

4.3. Biofilm Formation in Bacterial Populations

The formation of biofilms occurs in sequential steps: first, the attachment of the bacteria to the tissue or abiotic surface and second, the proliferation and accumulation of multilayer cell clusters with intercellular adhesion which relies on the production of extracellular matrix [69].

Three main components may influence biofilm formation: the bacterial cells, the attachment surface, and the surrounding environment [70]. The adhesion of bacterial cells, the first step of biofilm formation, is mostly influenced by the physicochemical properties of the cells' surface, if an abiotic surface is considered [71]. In biotic surfaces, the attachment is mainly dependent of specific interactions between bacterial adhesins and the molecules or receptors on the substrate surface [72]. The surface properties such as cleanability, disinfectability, roughness and wettability are important factors that affect the biofilm formation. Nevertheless, environmental factors such as hydrodynamics of the fluid, pH, temperature, osmolarity, oxigen gradients, nutrient composition, and the presence of other bacteria play important roles in the biofilm formation [73 - 76]. Combination of these factors strongly determines the microbial behavior to form a biofilm [75, 77].

After initial adhesion, bacteria multiply and form a multilayer cell cluster which involves intercellular adherence and production of polysaccharides. For example, in *Staphylococcus* genus, the polysaccharide intercellular adhesin (PIA) is a polysaccharide adhesin with an important role in the accumulation and aggregation of cells, being the *ica*ADBC operon involved in the synthesis of this polysaccharide [78].

Growth of the attached bacteria and the production of extracellular material like PIA lead to the formation of high density bacterial aggregates enclosed in an extracellular matrix, which characterizes mature biofilms [79].

Mature biofilms can undergo a detachment process due to quorum-sensing signalling and hydrodynamic mechanical forces, leading to the release of planktonic bacteria. These released bacteria may colonize different surfaces and start producing another biofilm matrix [72]. After biofilms have formed, they are difficult to be eradicated resulting in persistent microbial populations with increased aptitude to persist in the environment [80]. Furthermore, the presence of mixed bacterial populations within a biofilm is more common, increasing the difficulty in eradicating the biofilm [81]. The increased resistance due to biofilm formation may lead to an inefficient sanitizing regime, since bacteria within a biofilm are more resistant to biocides than their planktonic phenotype [80, 82].

4.4. Extrinsic Resistance Mechanisms

Acquired resistance to biocides can emerge by a mutation or acquisition of genetic material in the form of plasmids, transposons or integrons, as it occur with antibiotics [3, 83]. Several studies reported that biofilm constitute a good environment for the transfer of genetic material [84, 85], and that the development of stress in biofilms leads to different genes response and double-stranded DNA breaks causing the emergence of biocide resistance variants [86]. For example, in *E. coli* and *P. aeruginosa*, this response may afford protection against the activity of oxidizing agents [87, 88]. Another possible mechanism to explain bacterial biocide resistance is the up-regulation or induction of genes coding to multidrug efflux pumps in biofilms. Efflux pumps are systems that allow the bacterial survival by riding off toxic molecules [3].

The key role of QACs in health and food chain made their resistance mechanisms among the best studied, and several QAC efflux systems have been discovered in Gram-positive bacteria (*qac*A; *qac*B, and *smr*) and in Gram-negative bacteria (*emr*E; *qac*E, and *qac*E1) [89]. A plasmid carrying the *qac*A gene in *S. aureus,* also carries genes for resistance to trimethoprim (*drf*A) and aminoglycosides (*aac*A-*aac*D) [90], and is a well-known system responsible for its high level of resistance to QAC and cationic biocides [91].

Another *S. aureus* efflux gene, not located in a plasmid but residing on the chromosome, is the *nor*A gene, which is the primary determinant of fluoroquinolone efflux and clinically significant fluoroquinolone resistance [92, 93]. Among others, a plasmid-encoded silver efflux system was described in *Salmonella* typhimurium, and the *mex* efflux systems provide resistance to triclosan *in vitro* in *P. aeruginosa*, [93, 94].

Resistance to biocides can also be achieved by its inactivation or the inactivation of its toxic derivatives. For example, the resistance to oxidants chlorine, hydrogen peroxide and peracetic acid, can be obtained by overproduction of catalases or by induction of the *oxy*R or *sox*RS regulons in *E. coli*, all defense systems that lead to reduction of radical-mediated damage [90]. Biocides can affect multiple cellular components, therefore target site mutations are rare in biocide-resistant organisms

[22]. One exception is the triclosan resistance in *E. coli*, which is effected by mutations in the *fab*I gene encoding the enoyl-acyl carrier protein reductase of fatty acid biosynthesis [59].

Currently, one topic of discussion in the literature is the potential for biocide co-resistance and cross-resistance to clinically important antibiotics. Co-resistance occurs when the resistant phenotypes genes are located together on a mobile genetic element such as a plasmid, transposon, or integrin. Cross-resistance can occur when different antimicrobial agents have the same target or share a common route to achieve the same target. The result is the same, because the development of resistance to one antibacterial agent is accompanied by the appearance of resistance to another agent [90]. An example is triclosan and isoniazid which share a common target in *Mycobacterium tuberculosis* and *Mycobacterium smegmatis* [8].Another example is the expression of the *oxy*R regulon genes that results in cross-resistance between hydrogen peroxide, hypochlorous acid, and multiple antibiotics [95].

The reasonable use of the available antimicrobials is crucial to limit co- and cross-resistance, not only in the clinical context.

5. EUROPEAN REGULATION FOR BIOCIDE PRODUCTS

Biocides may pose risks to the ecosystem including humans, animals and the environment. To overcome this problem a new European directive was approved which came into force on the 1st of September 2013, this new directive aiming to provide a high level of protection for humans, animals and the environment repealed, the previous directive 98/8/EC [96, 97].

The objective of the new Regulation No 528/2012 is to develop the functioning of the European Union (EU) internal market in biocides ensuring high standards of environmental and human health safety. This Regulation also resolves several weaknesses that were identified during the 11 years of implementation of the Directive 98/8/EC.

The new text have main improvements simplifying the requirements for approving active substances, authorizing products and also applying a sharper approach to animal testing by encouraging and making data sharing compulsory. This regulation brings also a major innovation for the EU internal market with the creation of a Union authorization for biocidal products allowing the industry to directly place their products on the entire Union market [98].

The Regulation No 528/2012 applies without prejudice to a previous Regulation No 1907/2006 concerning the Registration, Evaluation, Authorization and Restriction of Chemicals (REACH) which was established by the European Chemicals Agency (ECHA). This agency manages the technical, scientific and administrative aspects of REACH [97, 99].

The main objective of REACH is to provide a high level of protection for human health and the environment, while maintaining the competitiveness and innovation of EU industry. REACH provides a single regulatory framework for the control of chemicals, replacing the previous controls, and ensures that information on the properties of chemicals is transmitted down the supply chain, thus enabling them to be safely handled [98 - 100].

The European Chemicals Agency is the regulatory authority that implements the EU's chemicals legislation for the benefit of both human and animal health, as well as the environment. It also promotes the innovation and competitiveness within the EU. ECHA helps companies to act in accordance with the legislation, improve the safe use of chemicals, provides information on chemicals and addresses chemicals of concern [100].

ECHA will provide both technical and scientific knowledge to the Member States under this new Regulation. ECHA will in particular be responsible for the assessment of applications for the Union authorization of biocidal products. It is expected that the total number of opinions to be delivered by ECHA will increase from 80 in 2014 to 300 in 2020 leading to major cost saving for the industry of €2.7 billion over a period of 10 years [98, 100].

CONCLUSION

In recent times, the global market for biocides has increased significantly and is expected to continue to grow. When used appropriately, these compounds can play an important role in controlling infection and/or preventing the dissemination of pathogenic microorganisms. However, further investigation is needed to gain a better understanding of the survival capabilities of these microorganisms under biocide pressure, as well as studies enlightening the mechanisms of biocide resistance.

Biocides are essential in preventing and controlling infections, especially in the healthcare industry. An adequate and rational use of these substances with or without other antimicrobial compounds like antibiotics may prevent the decrease on therapeutic options against infections.

CONFLICT OF INTEREST

The author(s) confirm that this chapter contents have no conflict of interest.

ABBREVIATIONS

AgSD	Silver Sulfadiazine
CRA	Chlorine Releasing Agent
DNA	Deoxyribonucleic acid
ECHA	European Chemicals Agency
EU	European Union
H_2O_2	Hydrogen Peroxide
IRA	Iodine Releasing Agent
MRSA	Methicillin-Resistant *Staphylococcus aureus*
PAA	Peracetic Acid
PIA	Polysaccharide Intercellular Adhesin
QAC	Quaternary Ammonium Compounds
REACH	Registration, Evaluation, Authorisation and Restriction of Chemicals
RNA	Ribonucleic acid

ACKNOWLEDGEMENTS

This study was conducted with the financial support of the Interdisciplinary Centre of Research in Animal Health (CIISA) from the Faculty of Veterinary Medicine from the University of Lisbon. Project funded by PEst-OE/AGR/U10276/2014. Rui Seixas (SFRH/BD/75836/2011) and Carla Mottola

(SFRH/BD/72872/2010) hold a PhD fellowship from Foundation for Science and Technology (FCT), Portugal.

REFERENCES

[1] Baltzer SA, Brown MH. Antimicrobial peptides: promising alternatives to conventional antibiotics. J Mol Microbiol Biotechnol 2011; 20(4): 228-35.
[http://dx.doi.org/10.1159/000331009] [PMID: 21894027]

[2] Falagas ME, Grammatikos AP, Michalopoulos A. Potential of old-generation antibiotics to address current need for new antibiotics. Expert Rev Anti Infect Ther 2008; 6(5): 593-600.
[http://dx.doi.org/10.1159/000331009] [PMID: 21894027]

[3] Bridier A, Dubois-Brissonnet F, Greub G, *et al.* Dynamics of the action of biocides in *Pseudomonas aeruginosa* biofilms. Antimicrob Agents Chemother 2011; 55(6): 2648-54.

[4] Webb R, Philips A, Speare R, *et al.* Controlling wildlife fungal disease spread: *in vitro* efficacy of disinfectants against *Batrachochytrium dendrobatidis* and *Mucor amphibiorum.* Dis Aquat Organ 2012; 99(2): 119-25.
[http://dx.doi.org/10.3354/dao02461] [PMID: 22691980]

[5] Srinivas A, Kaman L, Raj P, *et al.* Comparison of the efficacy of chlorhexidine gluconate *versus* povidone iodine as preoperative skin preparation for the prevention of surgical site infections in clean-contaminated upper abdominal surgeries. Surg Today 2014.
[http://dx.doi.org/10.1007/s00595-014-1078-y] [PMID: 25381486]

[6] Wittmer IK, Bader HP, Scheidegger R, *et al.* Significance of urban and agricultural land use for biocide and pesticide dynamics in surface waters. Water Res 2010; 44(9): 2850-62.
[http://dx.doi.org/10.1016/j.watres.2010.01.030] [PMID: 20188390]

[7] Hugo WB. Early studies in the evaluation of disinfectants. J Antimicrob Chemother 1978; 4(6): 489-94.
[http://dx.doi.org/10.1093/jac/4.6.489] [PMID: 361678]

[8] Russell AD. Introduction of biocides into clinical practice and the impact on antibiotic-resistant bacteria. J Appl Microbiol 2002; 92 (Suppl.): 121S-35S.
[http://dx.doi.org/10.1046/j.1365-2672.92.5s1.12.x] [PMID: 12000621]

[9] Gerba CP. Quaternary Ammonium Biocides: Efficacy in Application. Appl Environ Microbiol 2014. [PMID: 25362069]

[10] Jenkins S, Addy M, Wade W. The mechanism of action of chlorhexidine. A study of plaque growth on enamel inserts *in vivo.* J Clin Periodontol 1988; 15(7): 415-24.
[http://dx.doi.org/10.1111/j.1600-051X.1988.tb01595.x] [PMID: 3183067]

[11] World Health Organization WHO Model List of Essential Medicines , [October 2013];

[12] Rossmoore HW. Introduction to biocide use. In: Rossmoore HW, Ed. Glasgow: Blackie Academic and Professional 1995; pp. 1-8p.
[http://dx.doi.org/10.1007/978-94-011-1354-0_1]

[13] Chaidez C, Soto-Beltran M, Gerba CP, Tamimi AH. Reduction of risk of *Salmonella* infection from

kitchen cleaning clothes by use of sodium hypochlorite disinfectant cleaner. Lett Appl Microbiol 2014; 59(5): 487-92.
[http://dx.doi.org/10.1111/lam.12321] [PMID: 25163762]

[14] Oosterik LH, Peeters L, Mutuku I, *et al.* Susceptibility of avian pathogenic *Escherichia coli* from laying hens in Belgium to antibiotics and disinfectants and integron prevalence. Avian Dis 2014; 58(2): 271-8.
[http://dx.doi.org/10.1637/10680-100113-RegR] [PMID: 25055632]

[15] Petlin A, Schallom M, Prentice D, *et al.* Chlorhexidine gluconate bathing to reduce methicillin-resistant *Staphylococcus aureus* acquisition. Crit Care Nurse 2014; 34(5): 17-25.
[http://dx.doi.org/10.4037/ccn2014943] [PMID: 25274761]

[16] Sandle T, Vijayakumar R, Saleh Al Aboody M, *et al. In vitro* fungicidal activity of biocides against pharmaceutical environmental fungal isolates. J Appl Microbiol 2014; 117(5): 1267-73.
[http://dx.doi.org/10.1111/jam.12628] [PMID: 25155804]

[17] Tremblay YD, Caron V, Blondeau A, *et al.* Biofilm formation by coagulase-negative staphylococci: impact on the efficacy of antimicrobials and disinfectants commonly used on dairy farms. Vet Microbiol 2014; 172(3-4): 511-8.
[http://dx.doi.org/10.1016/j.vetmic.2014.06.007] [PMID: 24984943]

[18] Ortiz S, López V, Martínez-Suárez JV. The influence of subminimal inhibitory concentrations of benzalkonium chloride on biofilm formation by *Listeria monocytogenes.* Int J Food Microbiol 2014; 189: 106-12.
[http://dx.doi.org/10.1016/j.ijfoodmicro.2014.08.007] [PMID: 25136789]

[19] Khan R, Lee MH, Joo H, *et al.* Triclosan resistance in a bacterial fish pathogen *Aeromonas salmonicida* sub sp. *salmonicida* is mediated by an enoyl reductase fabV. J Microbiol Biotechnol 2014.
[PMID: 25370725]

[20] Fernández Fuentes MA, Abriouel H, Gadea R, *et al.* Synergistic activity of biocides and antibiotics on resistant bacteria from organically produced foods. Microb Drug Resist 2014; 20(5): 383-91.
[http://dx.doi.org/10.1089/mdr.2013.0234] [PMID: 24660956]

[21] Ye JZ, Yu X, Li XS, *et al.* [Antimicrobial resistance characteristics of and disinfectant-resistant gene distribution in *Staphylococcus aureus* isolates from male urogenital tract infection]. Zhonghua Nan Ke Xue 2014; 20(7): 630-6.
[PMID: 25095620]

[22] McDonnell G, Russell AD. Antiseptics and disinfectants: activity, action, and resistance. Clin Microbiol Rev 1999; 12(1): 147-79.
[PMID: 9880479]

[23] Russell AD. Similarities and differences in the responses of microorganisms to biocides. J Antimicrob Chemother 2003; 52(5): 750-63.
[http://dx.doi.org/10.1093/jac/dkg422] [PMID: 14519671]

[24] Lachapelle JM. Antiseptis and Disinfectants. In: Rustemeyer T, Elsner P, John SM, Meibach HI, Eds. Berlin, Heidelberg: Springer-Verlag 2012; pp. 385-95.

[25] McDonnell G, Burke P. Disinfection: is it time to reconsider Spaulding? J Hosp Infect 2011; 78(3): 163-70.
[http://dx.doi.org/10.1016/j.jhin.2011.05.002] [PMID: 21664533]

[26] Dvorak G. Disinfection 101 Iowa, USA Available from: http://www.cfsph.iastate.edu 2008. [cited 2014 December 3]

[27] Maillard JY. Bacterial target sites for biocide action J Appl Microbiol 2002; (92 Symposium Supplement)9: 16S-27S.

[28] Spaulding EH. Chemical disinfection and antisepsis in the hospital. J Hosp Res 1957; 9: 5-31.

[29] Favero MS, Bond WW. Chemical Disinfection of Medical and Surgical Materials. In: Block S, Ed. Disinfection, Sterilization, and preservation. (4th ed.), Philadelphia: Lea & Febiger 1991.

[30] Hugo WB. Disinfection mechanisms Principles and Practice of Disinfection, Preservation and Sterilization. (3rd ed.) Oxford: Blackwell Science 1999; pp. 258-83.

[31] Russell A, Chopra I. Understanding Antibacterial Action and Resistance. (2nd ed.), Chichester, UK: Ellis Horwood 1996.

[32] Maris P. Modes of action of disinfectants. Rev Sci Tech 1995; 14(1): 47-55.

[33] Morley PS. Biosecurity of veterinary practices. Vet Clin North Am Food Anim Pract 2002; 18(1): 133-155, vii.
[http://dx.doi.org/10.1016/S0749-0720(02)00009-9] [PMID: 12064165]

[34] Chalew TE, Halden RU. Environmental Exposure of Aquatic and Terrestrial Biota to Triclosan and Triclocarban. J Am Water Works Assoc 2009; 45(1): 4-13.
[http://dx.doi.org/10.1111/j.1752-1688.2008.00284.x] [PMID: 20046971]

[35] Orvos DR, Versteeg DJ, Inauen J, *et al.* Aquatic toxicity of triclosan. Environ Toxicol Chem 2002; 21(7): 1338-49.
[http://dx.doi.org/10.1002/etc.5620210703] [PMID: 12109732]

[36] Jeffrey DJ. Chemicals used as disinfectants: active ingredients and enhancing additives. Rev - Off Int Epizoot 1995; 14(1): 57-74.
[PMID: 7548972]

[37] Minghua M, Sun G. Antimicrobial cationic dyes: part 2-thermal and hydrolytic stability. Dyes Pigm 2004; 63(1): 39-49.
[http://dx.doi.org/10.1016/j.dyepig.2004.01.004]

[38] Nehrbass-Stuedli A, Boykin D, Tidwell RR, *et al.* Novel diamidines with activity against *Babesia divergens in vitro* and *Babesia microti in vivo.* Antimicrob Agents Chemother 2011; 55(7): 3439-45.
[http://dx.doi.org/10.1128/AAC.01482-10] [PMID: 21537025]

[39] Bell CA, Hall JE, Kyle DE, *et al.* Structure-activity relationships of analogs of pentamidine against *Plasmodium falciparum* and *Leishmania mexicana amazonensis.* Antimicrob Agents Chemother 1990; 34(7): 1381-6.
[http://dx.doi.org/10.1128/AAC.34.7.1381] [PMID: 2201254]

[40] Lanteri CA, Tidwell RR, Meshnick SR. The mitochondrion is a site of trypanocidal action of the aromatic diamidine DB75 in bloodstream forms of *Trypanosoma brucei.* Antimicrob Agents

Chemother 2008; 52(3): 875-82.
[http://dx.doi.org/10.1128/AAC.00642-07] [PMID: 18086841]

[41] Cai Z, Sun G. Antimicrobial finishing of acrilan fabrics with cetylpyridinium chloride. J Appl Polym Sci 2004; 94: 243-7.
[http://dx.doi.org/10.1002/app.20876]

[42] Silver S, Phung T, Silver G. Silver as biocides in burn and wound dressings and bacterial resistance to silver compounds. J Ind Microbiol Biotechnol 2006; 33(7): 627-34.
[http://dx.doi.org/10.1007/s10295-006-0139-7] [PMID: 16761169]

[43] Zarei M, Jamnejad A, Khajehali E. Antibacterial effect of silver nanoparticles against four foodborne pathogens. Jundishapur J Microbiol 2014; 7(1): e8720.
[http://dx.doi.org/10.5812/jjm.8720] [PMID: 25147658]

[44] Bocian E, Grzybowska W, Tyski S. Evaluation of mycobactericidal activity of selected chemical disinfectants and antiseptics according to European standards. Med Sci Monit 2014; 20: 666-73.
[http://dx.doi.org/10.12659/MSM.890175] [PMID: 24755666]

[45] Rutala WA, Weber DJ. Infection control: the role of disinfection and sterilization. J Hosp Infect 1999; 43 (Suppl.): S43-55.
[http://dx.doi.org/10.1016/S0195-6701(99)90065-8] [PMID: 10658758]

[46] Fulín P, Pokorný D, Slouf M, *et al.* [Effect of sterilisation with formaldehyde, gamma irradiation and ethylene oxide on the properties of polyethylene joint replacement components]. Acta Chir Orthop Traumatol Cech 2014; 81(1): 33-9.
[PMID: 24755055]

[47] Bounoure F, Fiquet H, Arnaud P. Comparison of hydrogen peroxide and peracetic acid as isolator sterilization agents in a hospital pharmacy. Am J Health Syst Pharm 2006; 63(5): 451-5.
[http://dx.doi.org/10.2146/ajhp040451] [PMID: 16484519]

[48] Bessems E. The effect of pratical condition on the efficacy of desinfectants. Int Biodeterior Biodegradation 1998; 41(3/4): 177-83.
[http://dx.doi.org/10.1016/S0964-8305(98)00022-5]

[49] Russell AD, McDonnell G. Concentration: a major factor in studying biocidal action. J Hosp Infect 2000; 44(1): 1-3.
[http://dx.doi.org/10.1053/jhin.1999.0654] [PMID: 10633046]

[50] Leggett MJ, McDonnell G, Denyer SP, *et al.* Bacterial spore structures and their protective role in biocide resistance. J Appl Microbiol 2012; 113(3): 485-98.
[http://dx.doi.org/10.1111/j.1365-2672.2012.05336.x] [PMID: 22574673]

[51] Mafu AA, Roy D, Goulet J, *et al.* Efficiency of sanitizing agents for destroying *Listeria monocytogenes* on contaminated surfaces. J Dairy Sci 1990; 73(12): 3428-32.
[http://dx.doi.org/10.3168/jds.S0022-0302(90)79040-6] [PMID: 2129169]

[52] Surdeau N, Laurent-Maquin D, Bouthors S, *et al.* Sensitivity of bacterial biofilms and planktonic cells to a new antimicrobial agent, Oxsil 320N. J Hosp Infect 2006; 62(4): 487-93.
[http://dx.doi.org/10.1016/j.jhin.2005.09.003] [PMID: 16478644]

[53] Belessi CE, Gounadaki AS, Psomas AN, *et al.* Efficiency of different sanitation methods on *Listeria*

monocytogenes biofilms formed under various environmental conditions. Int J Food Microbiol 2011; 145 (Suppl. 1): S46-52.
[http://dx.doi.org/10.1016/j.ijfoodmicro.2010.10.020] [PMID: 21093085]

[54] Wang H, Hu C, Hu X, *et al.* Effects of disinfectant and biofilm on the corrosion of cast iron pipes in a reclaimed water distribution system. Water Res 2012; 46(4): 1070-8.
[http://dx.doi.org/10.1016/j.watres.2011.12.001] [PMID: 22209261]

[55] Anderson RL, Carr JH, Bond WW, *et al.* Susceptibility of vancomycin-resistant enterococci to environmental disinfectants. Infect Control Hosp Epidemiol 1997; 18(3): 195-9.
[http://dx.doi.org/10.2307/30141982] [PMID: 9090548]

[56] Singh AV, Vyas V, Patil R, *et al.* Quantitative characterization of the influence of the nanoscale morphology of nanostructured surfaces on bacterial adhesion and biofilm formation. PLoS One 2011; 6(9): e25029.
[http://dx.doi.org/10.1371/journal.pone.0025029] [PMID: 21966403]

[57] Nilsson RE, Ross T, Bowman JP. Variability in biofilm production by *Listeria monocytogenes* correlated to strain origin and growth conditions. Int J Food Microbiol 2011; 150(1): 14-24.
[http://dx.doi.org/10.1016/j.ijfoodmicro.2011.07.012] [PMID: 21824672]

[58] Fraise AP. Susceptibility of antibiotic-resistant cocci to biocides. Symp Ser Soc Appl Microbiol 2002; 31(31): 158S-62S.
[http://dx.doi.org/10.1046/j.1365-2672.92.5s1.2.x] [PMID: 12481840]

[59] McMurry LM, Oethinger M, Levy SB. Overexpression of *mar*A, *sox*S, or *acr*AB produces resistance to triclosan in laboratory and clinical strains of *Escherichia coli.* FEMS Microbiol Lett 1998; 166(2): 305-9.
[http://dx.doi.org/10.1111/j.1574-6968.1998.tb13905.x] [PMID: 9770288]

[60] Langsrud S, Sundheim G. Factors contributing to the survival of poultry associated *Pseudomonas* spp. exposed to a quaternary ammonium compound. J Appl Microbiol 1997; 82(6): 705-12.
[http://dx.doi.org/10.1046/j.1365-2672.1997.00186.x] [PMID: 9202437]

[61] Barry CE III, Mdluli K. Drug sensitivity and environmental adaptation of mycobacterial cell wall components. Trends Microbiol 1996; 4(7): 275-81.
[http://dx.doi.org/10.1016/0966-842X(96)10031-7] [PMID: 8829336]

[62] Kolawole DO. Resistance mechanisms of mucoid-grown *Staphylococcus aureus* to the antibacterial action of some disinfectants and antiseptics. FEMS Microbiol Lett 1984; 25(2-3): 205-9.
[http://dx.doi.org/10.1111/j.1574-6968.1984.tb01457.x]

[63] Brown MR. The role of the cell envelope in resistance. In: Brown MR, Ed. Resistance of Pseudomonas aeruginosa. Chichester, England: John Wiley & Sons, Ltd. 1975; pp. 71-99.

[64] Götz F. Staphylococcus and biofilms. Mol Microbiol 2002; 43(6): 1367-78.
[http://dx.doi.org/10.1046/j.1365-2958.2002.02827.x] [PMID: 11952892]

[65] Flemming HC, Wingender J. The biofilm matrix. Nat Rev Microbiol 2010; 8(9): 623-33.
[PMID: 20676145]

[66] McSwain BS, Irvine RL, Hausner M, *et al.* Composition and distribution of extracellular polymeric substances in aerobic flocs and granular sludge. Appl Environ Microbiol 2005; 71(2): 1051-7.

[http://dx.doi.org/10.1128/AEM.71.2.1051-1057.2005] [PMID: 15691965]

[67] Ceri H, Olson ME, Stremick C, *et al.* The Calgary Biofilm Device: new technology for rapid determination of antibiotic susceptibilities of bacterial biofilms. J Clin Microbiol 1999; 37(6): 1771-6. [PMID: 10325322]

[68] Burmølle M, Webb JS, Rao D, *et al.* Enhanced biofilm formation and increased resistance to antimicrobial agents and bacterial invasion are caused by synergistic interactions in multispecies biofilms. Appl Environ Microbiol 2006; 72(6): 3916-23. [http://dx.doi.org/10.1128/AEM.03022-05] [PMID: 16751497]

[69] O'Toole G, Kaplan HB, Kolter R. Biofilm formation as microbial development. Annu Rev Microbiol 2000; 54: 49-79. [http://dx.doi.org/10.1146/annurev.micro.54.1.49] [PMID: 11018124]

[70] Van Houdt R, Michiels CW. Biofilm formation and the food industry, a focus on the bacterial outer surface. J Appl Microbiol 2010; 109(4): 1117-31. [http://dx.doi.org/10.1111/j.1365-2672.2010.04756.x] [PMID: 20522145]

[71] Oliveira R, Azeredo J, Teixeira P. The importance of physicochemical properties in biofilm formation and activity. London: IWA Publishing 2003.

[72] Kong KF, Vuong C, Otto M. *Staphylococcus* quorum sensing in biofilm formation and infection. Int J Med Microbiol 2006; 296(2-3): 133-9. [http://dx.doi.org/10.1016/j.ijmm.2006.01.042] [PMID: 16487744]

[73] Giaouris E, Chapot-Chartier MP, Briandet R. Surface physicochemical analysis of natural *Lactococcus lactis* strains reveals the existence of hydrophobic and low charged strains with altered adhesive properties. Int J Food Microbiol 2009; 131(1): 2-9. [http://dx.doi.org/10.1016/j.ijfoodmicro.2008.09.006] [PMID: 18954916]

[74] Gerbersdorf SU, Wieprecht S. Biostabilization of cohesive sediments: revisiting the role of abiotic conditions, physiology and diversity of microbes, polymeric secretion, and biofilm architecture. Geobiology 2014. [PMID: 25345370]

[75] Seixas R, Gabriel M, Machado J, *et al.* Effect of simulated gastrointestinal conditions on biofilm formation by Salmonella. 4,[5],12:i. Sci World Jr 2014. 2014 153956

[76] Stepanović S, Cirkovic I, Mijac V, *et al.* Influence of the incubation temperature, atmosphere and dynamic conditions on biofilm formation by *Salmonella* spp. Food Microbiol 2003; 20(3): 339-43. [http://dx.doi.org/10.1016/S0740-0020(02)00123-5]

[77] Goller CC, Romeo T. Environmental influences on biofilm development. In: Romeo T, Ed. Current Topics in Microbiology and Immunology Bacterial Biofilms 322. Gernamy: Springer-Verlag 2008; pp. 37-66.

[78] Cramton SE, Gerke C, Schnell NF, *et al.* The intercellular adhesion (*ica*) locus is present in *Staphylococcus aureus* and is required for biofilm formation. Infect Immun 1999; 67(10): 5427-33. [PMID: 10496925]

[79] Wilson M. Bacterial biofilms and human disease. Sci Prog 2001; 84(Pt 3): 235-54. [http://dx.doi.org/10.3184/003685001783238998] [PMID: 11732158]

[80] Seixas R, Machado J, Bernardo F, *et al.* Biofilm formation by *Salmonella enterica* serovar 1,4,[5],12:i:- Portuguese isolates: a phenotypic, genotypic, and socio-geographic analysis. Curr Microbiol 2014; 68(5): 670-7.
[http://dx.doi.org/10.1007/s00284-014-0523-x] [PMID: 24463530]

[81] Yang L, Liu Y, Wu H, *et al.* Current understanding of multi-species biofilms. Int J Oral Sci 2011; 3(2): 74-81.
[http://dx.doi.org/10.4248/IJOS11027] [PMID: 21485311]

[82] Carpentier B, Cerf O. Biofilms and their consequences, with particular reference to hygiene in the food industry. J Appl Bacteriol 1993; 75(6): 499-511.
[http://dx.doi.org/10.1111/j.1365-2672.1993.tb01587.x] [PMID: 8294303]

[83] Hannan S, Ready D, Jasni AS, *et al.* Transfer of antibiotic resistance by transformation with eDNA within oral biofilms. FEMS Immunol Med Microbiol 2010; 59(3): 345-9.
[PMID: 20337719]

[84] Nguyen KT, Piastro K, Gray TA, *et al.* Mycobacterial biofilms facilitate horizontal DNA transfer between strains of *Mycobacterium smegmatis.* J Bacteriol 2010; 192(19): 5134-42.
[http://dx.doi.org/10.1128/JB.00650-10] [PMID: 20675473]

[85] Ando T, Itakura S, Uchii K, *et al.* Horizontal transfer of non-conjugative plasmid in colony biofilm of *Escherichia coli* on food-based media. World J Microbiol Biotechnol 2009; 25: 1865-9.
[http://dx.doi.org/10.1007/s11274-009-0070-y]

[86] Boles BR, Singh PK. Endogenous oxidative stress produces diversity and adaptability in biofilm communities. Proc Natl Acad Sci USA 2008; 105(34): 12503-8.
[http://dx.doi.org/10.1073/pnas.0801499105] [PMID: 18719125]

[87] Sauer K, Camper AK, Ehrlich GD, *et al. Pseudomonas aeruginosa* displays multiple phenotypes during development as a biofilm. J Bacteriol 2002; 184(4): 1140-54.
[http://dx.doi.org/10.1128/jb.184.4.1140-1154.2002] [PMID: 11807075]

[88] Ren D, Bedzyk LA, Thomas SM, *et al.* Gene expression in *Escherichia coli* biofilms. Appl Microbiol Biotechnol 2004; 64(4): 515-24.
[http://dx.doi.org/10.1007/s00253-003-1517-y] [PMID: 14727089]

[89] Chapman JS. Biocide resistance mechanisms. Int Biod Biodeg 2003; 51(2): 133-8.
[http://dx.doi.org/10.1016/S0964-8305(02)00097-5]

[90] Chapman JS. Disinfectant resistance mechanisms, cross-resistance, and co-resistance. Int Biod Biodeg 2003; 51(4): 271-6.
[http://dx.doi.org/10.1016/S0964-8305(03)00044-1]

[91] Smith K, Gemmell CG, Hunter IS. The association between biocide tolerance and the presence or absence of qac genes among hospital-acquired and community-acquired MRSA isolates. J Antimicrob Chemother 2008; 61(1): 78-84.
[http://dx.doi.org/10.1093/jac/dkm395] [PMID: 17981834]

[92] Poole K. Efflux-mediated resistance to fluoroquinolones in gram-negative bacteria. Antimicrob Agents Chemother 2000; 44(9): 2233-41.
[http://dx.doi.org/10.1128/AAC.44.9.2233-2241.2000] [PMID: 10952561]

[93] Poole K. Mechanisms of bacterial biocide and antibiotic resistance. Symp Ser Soc Appl Microbiol 2002; (31): 55S-64S.
[http://dx.doi.org/10.1046/j.1365-2672.92.5s1.8.x] [PMID: 12481829]

[94] Gupta A, Matsui K, Lo JF, *et al.* Molecular basis for resistance to silver cations in *Salmonella.* Nat Med 1999; 5(2): 183-8.
[http://dx.doi.org/10.1038/5545] [PMID: 9930866]

[95] Greenberg JT, Demple B. A global response induced in *Escherichia coli* by redox-cycling agents overlaps with that induced by peroxide stress. J Bacteriol 1989; 171(7): 3933-9.
[PMID: 2472381]

[96] European Chemicals Agency (ECHA) Understanding BPR Available from: https://echa.europa.eu/regulations/biocidal-products-regulation/understanding-bpr , [cited 2014 November 23];

[97] European Commission (EC) Regulation (EC) 1907/2006 of the European Parliament and of the Council of 18 December 2006 concerning the registration, evaluation, authorisation and restriction of chemicals (REACH), establishing a European Chemicals Agency, amending Directive 1999/45/EC and repealing Council Regulation (EEC) 793/93 and Commission Regulation (EC) No. 1488/94 as well as Council Directive 76/769/EEC and Commission Directives 91/155/EEC, 93/67/EEC, 93/105/EC and 2000/21/EC. Off J Eur Union L 2006; 396: 374-5.

[98] European Commission (EU) Regulation concerning the making available on the market and use of biocidal products Available from:http://ec.europa.eu/environment/chemicals/biocides/regulation/regulation_en.htm , 2014 [cited 2014 November 23];

[99] European Chemicals Agency (ECHA) Understanding REACH Available from: http://echa.europa.eu/web/guest/regulations/reach/understanding-reach , 2014 [cited 2014 November 23];

[100] European Chemicals Agency (ECHA) Working for the safe use of chemicals Available from: http://echa.europa.eu/documents/10162/13556/echa_general_leaflet_en.pdf , 2014 [cited 2014 November 23];

CHAPTER 9

Novel Therapeutic Strategies in Veterinary Medicine

Ana Duarte*, Luis Tavares

Centro de Investigação Interdisciplinar em Sanidade Animal (CIISA), Faculdade de Medicina Veterinária, Universidade de Lisboa, Avenida da Universidade Técnica, 1300-477 Lisboa, Portugal

Abstract: The use of antibiotics in animal production has led to the emergence of antibiotic resistant microorganisms with severe consequences in animal and human health. To counterbalance and substitute the use of antibiotics as antimicrobial agents, a comprehensive effort has been directed towards the development of new therapeutic strategies and Veterinary Medicine offers an appealing field for the implementation of such an effort. Probiotics and bacteriophages are two of the alternatives discussed.

Keywords: Antibiotic resistance bacteria, Bacteriophages, Food-borne pathogens, Microbiome, Phage therapy, Probiotics.

1. INTRODUCTION

The emergence of antibiotic resistant microorganisms, caused by an improper use of antibiotics is a major public health concern, pressing towards the development of new therapeutic strategies, which may counterbalance or substitute the use of antibiotics as antimicrobial agents.

Veterinary Medicine provides an appealing field for the use and development of novel therapeutic strategies due to the close link and sanitary impact of animal health in public health. Intensive animal production in highly stressful conditions increases the animals' susceptibility to microbiota imbalance increasing the risk of colonization by pathogenic microorganisms. In small animal practice there is a growing awareness regarding the development of antibiotic resistance

* **Corresponding author Ana Duarte:** Centro de Investigação Interdisciplinar em Sanidade Animal (CIISA), Faculdade de Medicina Veterinária, Universidade de Lisboa, Avenida da Universidade Técnica, 1300-477 Lisboa, Portugal; Email: anaduarte@fmv.ulisboa.pt

Manuela Oliveira and Isa Serrano (Eds)

mechanisms by potential zoonotic microorganisms.

This chapter focuses on two strategies: probiotics and bacteriophage therapy and their scope in Veterinary Medicine. Probiotics interact with the host microbiota, contributing to a higher health performance and disease control, both in production animals, and in small animal's medicine. Bacteriophages are bacteria virus, used for the control of bacteria pathogens. Both are used as biocontrol agents, although with different applications and scope.

1.1. Probiotics

It was Ilya Ilyich Metchnikoff (1845–1916) who introduced the concept that probiotics contributed positively for the enhancement of health [1]. Since then, the scientific basis of this concept have been thoroughly studied and the basic mechanisms by which probiotics contribute to the host homeostasis are beginning to be unraveled.

Probiotics are defined as "live microorganisms, when administered in adequate amounts confer a health benefit on the host" [2]. They provide a mean to restore the host microbiota, lost by death or excretion. Presently probiotics include viable bacteria, yeast and fungi, without adverse effects, contributing to the host health [3].

The mechanisms by which probiotics restore intestinal balance and subsequently prevent disease include: (i) stimulation of the host immune response; (ii) production of antimicrobial products; (iii) competitive exclusion of pathogenic microorganisms; and (iv) blockage of microbial toxins, complementing and/or substituting the role of the commensal microbiota [4, 5]. These actions complement each other, enhancing microbiota diversity, preventing intestinal colonization by pathogenic microorganisms, contributing to infection quell, and the recovery of homeostasis between the microbiome and the host [6].

Each organism has its own microbiome with specific microbial species and strains and critical differences between domestic animals and human are easily identified, due to the high physiological diversity between animal species. The present knowledge supporting the use of probiotics in Human and Veterinary Medicine is mostly based on animal models and *in vitro* studies, although a substantial effort is being made to characterize which microbial species and strains should be used in each animal species and in which situation [6].

In 1989, Fuller recorded that a good probiotic should be: i) a strain capable of exerting a beneficial effect on the host animal, promoting increased growth or resistance to disease; ii) non-pathogenic and non-toxic; iii) present as viable cells, preferably in large numbers; iv) capable of surviving and metabolizing in the gut environment; and v) stable and capable of remaining viable for long periods under storage and field conditions [7]. Additionally probiotics should not acquire antibiotic resistance genes and should not recombine with pathogenic microorganisms [8]. This definition is still updated, but nevertheless open to the inclusion of new features according to recently gathered information on probiotics action.

In Veterinary Medicine, probiotics are mainly intestinal probiotics and are used as growth promotors in farm animals, improving food digestibility and also to stabilize gastro-intestinal microbiota and prevent intestinal infections [9]. In small animals, in gastroenterology due to their action as microbiota stabilizers.

probiotics improve nutrition and are also used

1.2. Chicken and Turkeys

In healthy poultry, a balance amid the intestinal microbiome persists. Whenever there's a change in environmental conditions leading to stress, a shift between microbial populations occurs, with a decrease in favorable species, mainly lactobacilli, and an increase in detrimental microbial species, with a direct consequence on poultry health and eventually leading to disease. Due to husbandry methods, young chickens are deprived from their parents being unable to receive gut flora from their mothers' feces, making probiotic supplementation in these animals particularly beneficial, due to its direct effect on the gut microbiome composition and balance [10].

In broiler nutrition, the most current microbial species used as probiotics are highly diversified and include *Lactobacillus*, *Streptococcus*, *Bacillus*, *Enterococcus*, *Pediococcus* and *Bifidobacterium* genus, *Escherichia coli*, yeasts

(*Saccharomyces cerevisiae*, *Candida*) and molds (*Aspergillus*) [3, 8, 10]. Probiotics may have one or more microorganisms, preferably host-specific, *Lactobacilli* being the most common, owing to their ability to colonize the intestinal epithelium.

The mechanisms by which probiotics hinder pathogenic bacteria, contributing to the decrease of disease includes: competitive exclusion and antagonism of pathogenic microorganisms; metabolism changes; enhancement of food digestibility and immunological stimulation. The competitive exclusion approach has been long used for the control of endemic and zoonotic microorganisms in poultry, including *Salmonella* and *Campylobacter* [10]. The prevention of pathogen colonization of the intestinal habitat is due to the previous establishment of a beneficial microorganism due to the completion of adherence sites in the intestinal mucosa [11]. Also probiotics produce antimicrobial compounds (bacteriocins, colicins, lactic and acetic acid, among others) with an inhibitory effect on pathogens growth [8]. Anti-enterotoxins neutralize the toxicity of pathogenic enterotoxins and contribute for intestinal balance/recovery. The use of cultured microorganism from the digestive tract of adult chickens, as feeding supplement for newly hatch chicks has proven effective for the control of *Salmonella enteriditis* and paratyphoid salmonellas but also pathogenic *Escherichia coli*, *Clostridium perfringens*, *Listeria monocytogenes*, *Yersinia enterocolitica*, *Staphylococcus aureus*, *Campylobacter jejuni*, *Candida albicans* and coccidian [8, 11 - 15]. Preselected enteric microbiota seems to enhance its efficacy towards *Campylobacter* [14]. The exact mixture of microorganisms remains uncertain but although several mixtures, including 10 to 50 microbial species are commercially available, none has proven as effective as raw intestinal material [10]. Probiotic supplements besides being effective in chickens have also proved their usefulness in turkey, quail and pheasant and their effectiveness is not affected by breed, strain or sex; however antibiotics, stress, disease and husbandry errors may disturb it [13, 16].

To further complement the mechanism of competitive exclusion of probiotics towards microbial pathogens, a direct stimulation of the immune systems is also evident. The immune stimulation achieved by probiotics, is directed to the cellular (Th1) and humoral (Th2) response [3, 8]. *Lactobacilli* have a positive effect on the expression of the Th2 profile cytokines responsible for B cells stimulation and antibody production [17, 18]. Probiotics supplementation is related to higher concentration of IgA, IgM and IgG producing cells and higher antibody titers against viral infections [19, 20].

The use of probiotics in chickens is related to increased growth rate, most probably due to metabolic changes in the intestinal environment, delivering enzymes and other compounds, thus increasing food digestibility [3, 13, 15, 21], with a positive impact on the feed conversion rate. Probiotics may produce anticholesterolemic and antilipidemic factors; synthetize amino-acids (lysine) and B vitamins; may change the gastrointestinal pH, optimizing the role of intestinal enzymes and, whereas pathogens require high protein concentration, probiotics, mainly lactobacillus, require carbohydrates [9]. In hens probiotic feeding increases the egg weight and shell thickness and its use in stress periods improves the performances of birds laying eggs [9, 15, 22].

1.3. Ruminants

The rumen is the most important digestive compartment in ruminants, with an extremely complex microbial ecosystem, including anaerobic bacteria, protozoa, and fungi among others. In these animals' probiotics, live yeast (*Saccharomyces cerevisiae*) as well as filamentous fungus, are frequently used, to enhance animal performance [9, 23]. Among the beneficial effects in dairy cattle, increase of dry matter intake and milk production is referred [24]. In beef cattle daily probiotic supplementation has also been reported to improve growth parameters. These positive effects were associated to an enhancement of fiber digestibility and higher microbial density in the rumen [25]. Yeast supplementation has also a stabilizing effect on the pH of the rumen, reducing the risk of acidosis, a metabolic disorder frequent in dairy and beef cattle, caused by an increase of lactic acid load. Live yeasts can influence the microbial production and uptake of lactate [26]; and their growth promote ruminal protozoa population, competing with lactate producing bacteria [27]. In calves, where rumen is still immature, bacterial and yeast probiotics aim the small intestine, and are frequently used to control pathogen colonization, stabilizing the microbiota [23].

Probiotics in cattle also contribute to reduce the load of zoonotic bacterial pathogens in ruminants [23]. Direct supplementation of bacterial probiotics, such as *Lactobacillus acidophilus* has shown an inhibitory effect on *Escherichia coli* O157:H7 and *Salmonella* shedding in beef cattle [28, 29].

Besides the gastrointestinal purpose, a different approach for bacterial probiotics in cattle is related to the prevention of bacterial infectious diseases, such as bovine mastitis and postpartum infections. The treatment in usually based on the use of antibiotics with several associated risks, including the development of antibiotic resistance and residues in the milk and meat. Lactic acid bacteria (LAB) and also *Bacilli*, are part of the microbiota of the teat and vaginal canal of cows and their usefulness in this context has been assessed [30 - 32]. Autochthonous strains act by a competitive exclusion mechanism, which includes epithelial cell adhesion and pathogen inhibition, producing antagonistic compounds such as lactic acid, H_2O_2 and bacteriocins, already characterized in LAB strains from the vagina and from LAB and Bacilli strains from the teat canal of cattle [30, 33, 34]. These strains are efficient against *Staphylococcus aureus*, a major pathogen frequently related to the development of bovine mastitis and also postpartum complications. When duly evaluated they could be used directly in the utter skin or in the vaginal canal [32]. Although further characterization is required to assess probiotics functionality and safety in this context, their use would allow the development of new control strategies against important bacterial pathogens, improving animal health and reducing economic costs.

1.4. Horses

In horses, the colon and cecum are the most important fermentative compartments, equivalent to the ruminant's rumen, providing a highly complex ecosystem inhabited by a diversified microbiota, including bacteria, protozoa and fungi. Ecosystem shifts, which imply dysbiosis have important consequences in the animal health and are currently associated with acute colitis, equine grass sickness, and laminitis [35, 36].

Although an overall picture is difficult to apprehend the major representative microbial phyla in healthy horses include *Bacteroidetes*, *Proteobacteria*,

Verrucomicrobia, *Actinobacteria*, and *Spirochaetes*; *Clostridiales* and *Lactobacillales*, with individual variations between animals [4]. The use of probiotics in horses aims disease control/prevention. *Lactobacillus*, *Bifido-bacterium* and enterococci, although not included in the most representative bacteria in the equine colon, are the most common bacteria used as probiotics, due to their colonization and survival characteristics, within the digestive tract [4], although with intra bacterial strains variations. A thorough characterization of the "normal" microbiome will definitely contribute to a robust use of probiotics in these animals.

Another aspect to bear in mind is the possibility of horizontal transfer of antibiotic resistance genes between probiotic strains and in-house microbiota, already confirmed for *Enterococci* and *Lactobacilli* strains by *in vitro* and *in vivo* studies, further justifying the need to thoroughly assess potential probiotics strains [37, 38].

Data on probiotics use on diseased horses is often conflicting. Administration of *Saccharomyces boulardii* to horses with acute enterocolitis, helped to reduce the severity of symptoms [39]. Also *Saccharomyces boulardii* was able to survive in the intestinal tract of horses with intestinal illness, although without any impact on the diarrhea duration [40]. *Lactobacillus pentosus* WE7 was evaluated for the prevention of neonatal diarrhea in foals, and unexpectedly led to a substantial exacerbation of symptoms [41]. On the other hand the use of a mixture of five equine Lactobacilli strains (*Lactobacillus salivarius* YIT 0479, *L. reuteri* YIT 0480, *L. crispatus* YIT 0481, *L. johnsonii* YIT 0482 and *L. equi* YIT 0483) increased the body weight and decreased diarrhea incidence in foals [42]. Regarding the use of probiotics to control Salmonella shedding in horses results have not been promising [4].

Overall, the reports about the use probiotics in horses justify a higher scientific investment to fully scrutinize which microorganisms and strains to be used and to further investigate the mechanisms through which probiotics exert their action, in order to choose the most effective.

1.5. Swine

Due to the ban of antibiotics as feed supplements, imposed by the European Union in 2006 in the swine industry, the use of probiotics stands as one among the viable alternatives to counteract the drawback of antibiotics removal, namely increased morbidity and mortality.

The most commonly used probiotics in swine include yeasts and bacteria aiming the caecum and colon, modulating host and microbiota interactions [43]. Although several studies point to beneficial effects regarding the use of probiotics in swine, a meaningful evaluation is hard to accomplish, regarding the microbial strains to use, the dose and treatment duration, the production systems and the animals' genotype. Nevertheless, probiotics have proven their usefulness. Feed supplementation with *Enterococcus faecium* DSM 7134 to gestating sows contributed to higher feed intake, weight performance and litter size [44]. Similarly gestating sows and their respective piglets fed with *Bacillus cereus* var. *toyoi*, nursed more piglets. Also the piglets recorded a lower incidence of liquid feces and post weaning diarrhea [45].

In swine the weaning period implies a highly complex set of physiological changes in the digestive compartments, favoring intestinal disorders, leading frequently to diarrhea and impaired growth performance [46]. Probiotics use in this period is recommended to prevent pathogen colonization and promote the gut colonization with beneficial microbiota, reducing the risk of necrotizing enterocolitis, a serious gastrointestinal disease [47]. Probiotic treatment after weaning, with *Lactobacillus sobrius* DSM 16698, resulted in a decrease of enterotoxigenic *Escherichia coli* load in the ileum, frequently associated with post weaning diarrhea, and improved the daily weight gain [48]. Also, feeding of weaning pigs with two strains of *Lactobacillus plantarum* (DSMZ 8862 and 8866) contributed to increase the microbiota diversity in the gastro intestinal tract [49].

Live yeast supplementation presents an additional advantage due to their resistance inside the digestive canal. Feed supplementation with *Saccharomyces cerevisiae* ssp. *boulardii* to piglets after weaning contributed to a rapid renewal of the intestinal mucosal, safeguarding local resistance, preventing diarrhea and intestinal infections. Additionally, piglets showed a higher growth performance [50].

1.6. Aquaculture

In aquaculture, as with other animal production systems, the usage of antibiotics led to the development of antibiotic resistance bacteria with severe economic consequences besides the human health risk, justifying the design and implementation of control measures and the development of alternatives, in which probiotics are included.

In aquatic organisms the interdependence between the animal and the surrounding environment is stronger than in terrestrial animals. Therefore, pathogens which are able to remain in the environment are taken up regularly by the animal during osmoregulation and feeding [51]. The dominance between the external and internal microbiota on the animal microbiota is controlled in same species, as observed in the Atlantic halibut, but not in others as in bivalve larvae [52, 53]. Due to this unique relationship probiotics in aquaculture are envisaged as whole or components of microorganisms, beneficial to the host but also to the ecosystem microbial community [51].

The most common microorganisms used in aquaculture are considerably broader comparing to terrestrial animals, and include bacteriophages, Gram negative and Gram positive bacteria, microalgae and yeasts [54, 55]. The use of probiotics has been associated to increased feeding performance, as with *Enterococcus faecium*, *E. durans*, *Leuconostoc* sp., *Streptococcus* sp. I and *Streptococcus* sp. II in Nile tilapia (*Oreochromis niloticus*); *Shewanella putrefaciens* Pdp11 in Senegalese sole (*Solea senegalensis*); and *Bacillus* sp., *Pediococcus* sp., *Enterococcus* sp. and *Lactobacillus* in spin rainbow trout (*Oncorhynchus mykiss*), among others [56 - 58].

Probiotics use is also related to the development of disease resistance. The use of *Bacillus* spp. decreased the *Vibrio* counts in white shrimps (*Litopenaeus vannamei*) and in the water; *Pediococcus pentosaceus* strain 4012 conferred protection against photobacteriosis caused by *Photobacterium damselae* subsp. *Piscicida*, in cobia (*Rachycentron canadum*) [59]. In what concerns viral diseases, several probiotics have also antiviral effect. *Bacillus* MCCB101 and *Micrococcus* MCCB104 stimulate the expression of AMP genes pre- and post-challenge with white spot virus (WSV) a major shrimp pathogen [60]; *Pseudomonas* sp., *Vibrios* sp., *Aeromonas* sp. showed antiviral activity against infectious hematopoietic necrosis virus (IHNV) and *Oncorhynchus masou* virus (OMV), affecting *Salmonidae* [61].

The modes of action of probiotics in aquaculture include competitive exclusion, colonization of the digestive mucosal epithelium and the intestinal mucus [54]. Several publications thoroughly enumerate numerous examples of pathogenic bacteria, inhibited by different probiotics [54]. *Bacillus amyloliquefaciens* FPTB16 contributes for resistance against *Edwardsiella tarda* infection of catla (*Catla catla*); *Pediococcus acidilactici* improves shrimp (*Litopenaeus stylirostris*) culture affected by vibriosis and *Pediococcus pentosaceus* protects groupers (*Epinephelus* spp.) from photobacteriosis, caused by *Vibrio anguillarum* [62 - 64]. This protective action is due to a positive effect on the animal immunity, enhancing the innate, humoral and the cellular immunity [65 - 67].

1.7. Cat and Dog

The use of probiotics in small animal medicine, likewise to other animal species, aims the host gastrointestinal health. In the dog and cat the small intestine harbors a complex microbiota, although with physiological differences between the canine and the feline species. In cats anaerobic bacteria are present in proportions that are considered abnormal in dogs and humans [68].

Several studies address the beneficial effects of probiotics in small animal nutrition. Probiotic *Lactobacillus acidophilus* DSM13241 has proven its usefulness in cats, when added to dry food. A decrease in unfavorable intestinal bacterial markers (clostridia and enterococci) and an increase in beneficial lactobacilli were observed, suggesting a shift within the intestinal microbiota, during probiotic supplementation [68]. Hematological changes regarding lymphocytes, eosinophils and peripheral blood phagocytes populations were also suggestive of probiotic immunomodulation. Also the probiotic *Enterococcus faecium* SF68, probably due to its effect as an immunostimulant in cats, contributed to microbiota diversity and to the control of Feline Herpesvirus reactivation [69, 70].

Similarly to cats, *Enterococcus faecium* SF68 when used as food supplement in puppies enhanced the immune response, as observed by an increase of the mucosal humoral response and an increase of specific Canine Distemper Virus (CDV) antibodies (IgG), after vaccination against CDV [71]. Due to probiotics beneficial effect, s everal dog bacterial strains have been assessed *in vitro* and *in vivo*, regarding their resistance to the intestinal milieu, production of bacteriocins, adhesive capacity and antibiotic resistance. Besides *E. faecium*, lactic acid bacteria, including lactobacilli and bifidobacteria have demonstrated their ability to compete with bacterial pathogens with zoonotic potential as *Clostridium perfringens*, *Clostridium difficile*, *Staphylococcus intermedius* and *Salmonella Typhimurium* [71 - 73]. However, with respect to enterococci a positive cooperation was observed regarding *Campylobacter jejuni*, raising a safety concern in the evaluation of bacterial strains as probiotics.

In small animal medicine, several studies point to the effectiveness of probiotics in canine inflammatory bowel disease, due to the participation of microbiota in the pathogenesis of enteropathies. *Ex vivo* studies demonstrated a positive immunological modulation of lactobacilli cocktails towards intestinal inflammation [74, 75]. These findings were recently confirmed by *in vivo* studies, confirming a modulating effect of probiotics on the mucosal immune system, but also on the microbiota composition [76].

Probiotic microorganisms positively influence the host homeostasis, contributing to animal health. Although probiotic strains have been widely used and will continue to be, the relationships between them and the animal microbiome remains to be fully assessed. The current scientific investment in the characterization of human microbiome will hopefully be followed by a similar scientific effort on other terrestrial and aquatic organisms, allowing a more ecological perspective on microbial interactions, essential for a rational use of probiotics. In view of the growing public awareness towards the indiscriminate use of antibiotics and the emergence of bacteria with a broad spectrum of antibiotic resistance, probiotics offer a safer and sustainable alternative to improve animal health and growth performance and consequently contributing to improve human health.

2. BACTERIOPHAGES AND PHAGE THERAPY

Bacteriophages (phages) are viruses that infect bacteria. Due to their specificity, a given phage only infect the correspondent bacteria species/strain, which in view of the current concern on the development of bacterial antibiotic resistance makes them excellent tools to control bacteria.

Phages were viewed as bacterial killers, since the end of the XIX century and were used as therapeutic agents against bacterial infections until before the II World War, when the discovery of penicillin relegated their use in the Western World to background levels. Nevertheless phage therapy continued to be used in Eastern Europe. Presently, with the emergence of bacteria with a broad antibiotic resistance range, phage therapy reemerged in the West, and old and new applications are currently under study [77].

Phages remain within a susceptible bacteria population as lytic (virulents) or lysogenic (temperate). The typical replication cycle of a lytic phage includes cell entry, mediated by the attachment of phage particles to specific cell wall receptors, assuring specificity towards a given bacteria population, followed by genome replication inside the cell, synthesis of the viral capsid proteins and assembly of new particles, exiting the cell by membrane and cell wall rupture, thus inducing cell death. Within the context of phages as antimicrobial agents only virulent lytic phages are used [78]. As long as a susceptible bacteria population is available the lytic phage remains viable; afterwards, when living cells are no longer available, it will be naturally eliminated. Due to the high specificity of phage infection, commensal cells are not affected and they present low toxicity for the animal host. In fact, phages are the most numerous biochemical units present in our environment and our continuous exposure to them, certify their safety [79].

The main difference between lytic and temperate phages is that the later integrate their genome within the bacterial genome, as prophage, in the beginning of the replication cycle, silencing its presence within the cell population. Excision of the prophage and completion of the replication cycle may occur and when it does, the new phage particles may include bacterial genomic fragments, participating on the horizontal gene transfer within bacteria populations, acting as genetic shuttles, delivering genetic information between bacterial cells, including antibiotic resistance genes and virulence genes [80]. This process is known as transduction and represents a driving force in bacterial evolution, together with transformation and conjugation the other two processes of genetic interchangeability in bacteria populations. Although temperate phages are not used as control or therapeutic agents against bacterial infections they can be used as genetic cassettes delivering "good" genes within a "bad" bacteria population, reverting antibiotic resistance genes, changing bacteria genotypes into friendly phenotypes [81]. On the other hand, temperate phages, persists in the bacteria population virtually undetectable, and as carriers of genetic information between cells, their presence imply the transference of potentially pathogenic information, conferring higher cell robustness, thus leading to more virulent bacteria populations [80].

2.1. Scope of Phage Therapy in Veterinary Sciences

In Veterinary the scope for using phages as antimicrobials is massive. Phage therapy has already been assessed for the control of food borne pathogens and as therapeutic agents towards pathogenic bacteria. In 2006, the US Food and Drug Administration approved the use of the first phage cocktail (ListShield™), effective against *Listeria monocytogenes* in food products [82]. Since then, numerous scientific papers report the research efforts to characterize and evaluate new phages effective against bacterial pathogens and new product are on the market, namely: ListexTM also against *Listeria monocytogenes*; Ecoshield™, targeting *Escherichia coli* O157:H7 in meat and vegetables; BacWash™ as bio sanitizer of *Salmonella* and *E. coli* O157:H7 for direct application in the animals among others [78].

Phages are effective in free range chickens as a prophylactic and therapeutic agent, for the control of *Salmonella* Enteriditis PT4 contamination and infection. Oral feeding of phages isolated from chickens reduced the bacterial load of the feces, contributing for a reduced contamination of birds and poultry products by *Salmonella enterica* Serovar Enteriditis [83, 84]. Application of a phage cocktail spray in the eggs surface to control the horizontal transmission of *Salmonella* Enteriditis, reduced the disease severity in chicks [85]. Phage therapy decreased colonization by *Campylobacter jejuni* and *Campylobacter coli*, both commensal bacteria in birds but causing campylobacteriosis in humans, when consumed from poultry meat products [86 - 88]. *Listeria monocytogenes* also has zoonotic potential and can be found in a wide variety of food products. Several phage cocktails are currently available for the reduction of *Listeria monocytogenes* load, from the surface of raw salmon fillets, smear-ripened cheese, broilers meat and also fresh fruits and juices, among others [89, 90]. The control of *Vibrio parahaemolyticus*, responsible for an important foodborne illness, through the ingestion of raw seafood mainly oysters was achieved by surface application of phages [91].

Phages are particularly appealing to complement and/or substitute antibiotic therapy. In aquaculture, a fast growing industry, several pathogenic microorganisms, including bacterial, viruses, fungus and parasites, pose serious health threats. As bacterial diseases are the main problem, the indiscriminate use of antibiotics led to the development of antibiotic resistance bacterial strains, posing a health risk, both for animals and consumers, making phages a viable and environmental friendly alternative to control fish pathogens. Phage therapy to control and treat pathogenic bacteria infection has been evaluated in multiple publications [92]. Control of *Pseudomonas aeruginosa*, in ulcerative lesions of catfish species (*Clarias gariepinus*) was achieved, using specific phages [93]. *Pseudomonas plecoglossicida* causes a hemorrhagic ascites infection in ayu (*Plecoglossus altivelis*). The use of phage cocktails demonstrated potential as an alternative therapeutic strategy to control and treat this bacterial infection in aquaculture [94]. Treatment and prevention of *Aeromonas hydrophila* infection, the etiological agent of hemorrhagic septicemia in fresh water and marine fish, was also successful, using phage therapy [95]. *Flavobacterium psychrophilum* affects cold water fishes, and is considered a serious threat to salmonids aquaculture. The use of phages as food additive has also shown promising results in the control of this bacterial infection in rainbow trout [96].

Poultry production systems also deal with similar problems concerning antibiotic resistance bacteria and the development of alternative control measures including phage therapy, is under study. *Escherichia coli* pose an important threat in poultry industry, causing high mortality in broilers. Phages specific for several *E. coli* serotypes have been identified and characterized, with encouraging results regarding their use as control agents of *E. coli* infection [97]. However, not all reports had promising results, showing that extrapolation of *in vitro* results to *in vivo* studies does not always prove successful [98]. Broilers, including chickens are natural reservoirs of *Salmonella*, including the major human pathogens *Salmonella enterica* serotypes Enteritidis, Hadar, and Typhimurium, causing severe enteric infections. Using phages isolated from farms, abattoirs and wastewater, the intestinal *Salmonella* colonization was reduced, limiting the transmission risk to humans [99, 100]. Besides *Salmonella*, broilers can also be colonized of *Campylobacter jejuni*. Oral administration to broiler chicken of phages specific for *Campylobacter* controlled the colonization rate, limiting the bacteria entry to the food chain [101].

In cattle, phage therapy is under assessment for the control of *Staphylo - coccus aureus* a major mastitis causing pathogen, owing to the high antibiotic resistance pattern of these bacteria. Several phages were isolated from sewage, waste water and from animals with mastitis and evaluated for therapeutic use, regarding their lytic potential and host range [102 - 104]. The use of Mycophages, phages specific against *Mycobacteria* species, has also shown promising results in the control of *Mycobacteria* in milk [105]. *E. coli* O157:H7 is a zoonotic bacterium, with severe pathogenic impact in humans. Cattle are often carriers, shedding the bacteria to the environment. Meat products can also be contaminated, due to fecal contamination during slaughter and carcass processing. The control of *E. coli* shedding by phage therapy was assessed, aiming a reduction in bacteria excretion, before slaughter. A significant decrease of *E. coli* O157:H7 load was achieved, making phage therapy a viable alternative to reduce transmission of zoonotic pathogens from farm to fork [106]. Similarly sheep are also reservoirs of *E. coli* O157:H7. Oral administration of phages resulted in a decrease shedding of bacteria, without a measurable effect on local microbiota, summing to the positive scientific reports on the use of phage therapy in the control of bacteria pathogens [107].

In swine, *Salmonella* causes morbidity and mortality, being an important foodborne pathogen. *Salmonella* shedding is enhanced during stressful conditions, including transportation, increasing the risk of contaminated end products, during carcass processing. Oral feeding of pigs with phage cocktails reduced *Salmonella* intestinal colonization and fecal load [108, 109]. Additionally, the use of phage cocktails isolated from environmental sources, including pig feces, specific against *Salmonella* Typhimurum, prove its effectiveness as a pig skin biosanitizer [110].

Biocontrol of bacterial pathogens by phage therapy, including food-borne pathogens and pathogens with clinical expression towards the animal host with or without zoonotic potential, implies a comprehensive approach. Phage cocktails should include lytic phages targeting specific bacteria. Due to the high specificity of the phage towards a given cell host, an accurate identification and characterization of the targeted bacteria should be performed, prior to choose the correct phage to be used against it [79]. Also to choose the right phage, the most probable source to find it would be closer to the pathogen source, including the animal itself, waste products as well as the environment. Due to the ubiquity of phages, virtually every place on earth would provide endless sources to find the correct phage.

Despite the vast and promising application fields of phage therapy in Veterinary Sciences, several drawbacks have to be addressed. Bacteria cells can become resistant to phage infection, as they have towards antibiotics, although at a lower rate [80]; the mechanism of phage resistance reside in the presence/absence of phage receptors in the cell surface and also on the characteristics of the bacteria cell wall [80]. Phages and bacteria are dynamic duets, with a complex set of interactions not fully understood, profoundly linked to evolutionary thrives. The development of phage resistant cells, will inevitably lead to the development of new phages, within the bacteria population, able to infect former resistant cells [78, 111]. One possible way to overcome this downside in phage therapy is to use phage cocktails, increasing the phage spectrum against different strains of the same pathogen ensuring cross-strain lytic activity [111].

The administration route should also be considered. Phages are used orally (water/feed), targeting the gastro intestinal organs or by spraying for topical or respiratory use by inhalation or nebulization. Phages are sensible to acidic pH [112]. The efficacy of gastro-intestinal delivery can be enhanced by increasing the stomach and upper small intestine pH, prior to phage uptake, or by protecting the phage suspension with biopolymers from the adverse gastro environment, guaranteeing phage activity [108, 113, 114].

Phage dispersal and the accessibility of the targeted bacteria are influenced by the matrix composition. Solid matrix hinders phage adsorption decreasing the efficacy of infection [78]. Phage biocontrol in foods is more effective in liquid medium than in solid surfaces [115]. The presence of protective compounds in the medium also influences phage adsorption: for example, the presence of immune factors in the milk hampers phage adsorption to *Staphylococcus aureus* [116].

Besides the above mentioned factors, the effectiveness of phage therapy must also consider the targeted pathogen concentration and accessibility, phage preparation and dose, and environmental conditions. Despite the promising results that phage therapy has already demonstrated, care should be taken when *in vitro* assessments are extrapolated for *in vivo* experiments.

At the present stage of antibiotic resistance, further research on phage therapy is both needed and increasingly challenging. Long live microbiologists!

CONCLUDING REMARKS

Probiotics and phage therapy are two of the most reported novel therapies in Veterinary Medicine. However, new approaches should be explored in this area including immunotherapy, antimicrobial peptides and natural compounds.

Due to the intimate link between animal and human health, the development of new preventive and therapeutic methods would greatly contribute to the control of antimicrobial resistance bacteria, avoiding further dissemination to humans and the environment.

CONFLICT OF INTEREST

The authors confirm that this chapter contents have no conflict of interest.

ACKNOWLEDGEMENTS

Declared none.

REFERENCES

[1] Mackowiak PA. Recycling metchnikoff: probiotics, the intestinal microbiome and the quest for long life. Front Public Health 2013; 1: 52. [http://dx.doi.org/10.3389/fpubh.2013.00052] [PMID: 24350221]

[2] Reid G. Probiotics to prevent the need for, and augment the use of, antibiotics. Can J Infect Dis Med Microbiol 2006; 17(5): 291-5. [PMID: 18382642]

[3] Lutful Kabir SM. The role of probiotics in the poultry industry. Int J Mol Sci 2009; 10(8): 3531-46. [http://dx.doi.org/10.3390/ijms10083531] [PMID: 20111681]

[4] Schoster A, Weese JS, Guardabassi L. Probiotic use in horses - what is the evidence for their clinical efficacy? J Vet Intern Med 2014; 28(6): 1640-52. [http://dx.doi.org/10.1111/jvim.12451] [PMID: 25231539]

[5] Wynn SG. Probiotics in veterinary practice. J Am Vet Med Assoc 2009; 234(5): 606-13. [http://dx.doi.org/10.2460/javma.234.5.606] [PMID: 19250038]

[6] Oelschlaeger TA. Mechanisms of probiotic actions - A review. Int J Med Microbiol 2010; 300(1): 57-62. [http://dx.doi.org/10.1016/j.ijmm.2009.08.005] [PMID: 19783474]

[7] Fuller R. Probiotics in man and animals. J Appl Bacteriol 1989; 66(5): 365-78. [http://dx.doi.org/10.1111/j.1365-2672.1989.tb05105.x] [PMID: 2666378]

[8] Dhama K, Verma V, Sawant PM, Tiwari R, Vaid RK, Chauhan RS. Applications of probiotics in poultry: enhancing immunity and beneficial effects on production performances and health - a review. J Immunol Immunopathol 2011; 13: 1-19.

[9] Ezema C. Probiotics in animal production: A review. J Vet Med Anim Health 2013; 5: 308-16.

[10] Fuller R. The chicken gut microflora and probiotic supplements. Jpn Poult Sci 2001; 38: 189-96. [http://dx.doi.org/10.2141/jpsa.38.189]

[11] Revolledo L, Ferreira AJ, Mead GC. Prospects in *Salmonella* control: competitive exclusion, probiotics, and enhancement of avian intestinal immunity. J Appl Poult Res 2006; 15: 341-51. [http://dx.doi.org/10.1093/japr/15.2.341]

[12] Rantala M, Nurmi E. Prevention of the growth of *Salmonella infantis* in chicks by the flora of the alimentary tract of chickens. Br Poult Sci 1973; 14(6): 627-30. [http://dx.doi.org/10.1080/00071667308416073] [PMID: 4759990]

[13] Schneitz C. Competitive exclusion in poultry—30 years of research. Food Contr 2005; 16: 657-67. [http://dx.doi.org/10.1016/j.foodcont.2004.06.002]

[14] Bhaskaran HP, Donoghue AM, Arsi K, *et al. In vitro* selection of enteric microflora for potential use as a competitive exclusion culture against *Campylobacter* in poultry. Int J Poult Sci 2011; 10: 940-5.

[http://dx.doi.org/10.3923/ijps.2011.940.945]

[15] Alloui M, Szczurek W, Świątkiewicz S. The Usefulness of Prebiotics and Probiotics in Modern Poultry Nutrition. Ann Anim Sci 2013; 13: 17-33.

[16] Schneitz C, Renney DJ. Effect of a commerical competitive exclusion product on the colonization of *Salmonella infantis* in day-old pheasant chicks. Avian Dis 2003; 47(4): 1448-51.
[http://dx.doi.org/10.1637/7031] [PMID: 14708995]

[17] Christensen HR, Frøkiaer H, Pestka JJ. Lactobacilli differentially modulate expression of cytokines and maturation surface markers in murine dendritic cells. J Immunol 2002; 168(1): 171-8.
[http://dx.doi.org/10.4049/jimmunol.168.1.171] [PMID: 11751960]

[18] Maassen CB, van Holten-Neelen C, Balk F, *et al.* Strain-dependent induction of cytokine profiles in the gut by orally administered *Lactobacillus* strains. Vaccine 2000; 18(23): 2613-23.
[http://dx.doi.org/10.1016/S0264-410X(99)00378-3] [PMID: 10775795]

[19] Yurong Y, Ruiping S, Shimin Z, Yibao J. Effect of probiotics on intestinal mucosal immunity and ultrastructure of cecal tonsils of chickens. Arch Anim Nutr 2005; 59(4): 237-46.
[http://dx.doi.org/10.1080/17450390500216928] [PMID: 16320812]

[20] Talebi A, Amirzadeh B, Mokhtari B, Gahri H. Effects of a multi-strain probiotic (PrimaLac) on performance and antibody responses to Newcastle disease virus and infectious bursal disease virus vaccination in broiler chickens. Avian Pathol 2008; 37(5): 509-12.
[http://dx.doi.org/10.1080/03079450802356995] [PMID: 18798026]

[21] Abu-Ruwaida AS, Husseini M, Banat IM. *Salmonella* exclusion in broiler chicks by the competitive action of adult gut microflora. Microbios 1995; 83(334): 59-69.
[PMID: 7476569]

[22] Zarei M, Ehsani M, Torki M. Dietary Inclusion of probiotics, prebiotics and symbiotic and evaluating performance of laying hens. Am J Agric Biol Sci 2011; 6: 249-55.
[http://dx.doi.org/10.3844/ajabssp.2011.249.255]

[23] Chaucheyras-Durand F, Durand H. Probiotics in animal nutrition and health. Benef Microbes 2010; 1(1): 3-9.
[http://dx.doi.org/10.3920/BM2008.1002] [PMID: 21840795]

[24] Jouany JP. Optimizing rumen functions in the close-up transition period and early lactation to drive dry matter intake and energy balance in cows. Anim Reprod Sci 2006; 96(3-4): 250-64.
[http://dx.doi.org/10.1016/j.anireprosci.2006.08.005] [PMID: 16996704]

[25] Guedes CM, Gonçalves D, Rodrigues MA, Dias-da-Silva A. Effects of a *Saccharomyces cerevisiae* yeast on ruminal fermentation and fiber degradation of maize silages in cows. Anim Feed Sci Technol 2008; 145: 27-40.
[http://dx.doi.org/10.1016/j.anifeedsci.2007.06.037]

[26] Rossi F, Luccia AD, Vincenti D, Cocconcelli PS. Effects of peptidic fractions from *Saccharomyces cerevisiae* culture on growth and metabolism of the ruminal bacteria *Megasphaera elsdenii.* Anim Res 2004; 53: 177-86.
[http://dx.doi.org/10.1051/animres:2004009]

[27] Brossard L, Chaucheyras-Durand F, Michalet-Doreau B, Martin C. Dose effect of live yeasts on rumen

microbial communities and fermentations during butyric latent acidosis in sheep: new type of interaction. Anim Sci 2006; 82: 829-36.
[http://dx.doi.org/10.1017/ASC200693]

[28] Tabe ES, Oloya J, Doetkott DK, Bauer ML, Gibbs PS, Khaitsa ML. Comparative effect of direct-fed microbials on fecal shedding of *Escherichia coli* O157:H7 and *Salmonella* in naturally infected feedlot cattle. J Food Prot 2008; 71(3): 539-44.
[PMID: 18389697]

[29] Stephens TP, Loneragan GH, Karunasena E, Brashears MM. Reduction of *Escherichia coli* O157 and *Salmonella* in feces and on hides of feedlot cattle using various doses of a direct-fed microbial. J Food Prot 2007; 70(10): 2386-91.
[PMID: 17969623]

[30] Espeche MC, Pellegrino M, Frola I, Larriestra A, Bogni C, Nader-Macías ME. Lactic acid bacteria from raw milk as potentially beneficial strains to prevent bovine mastitis. Anaerobe 2012; 18(1): 103-9.
[http://dx.doi.org/10.1016/j.anaerobe.2012.01.002] [PMID: 22261519]

[31] Bouchard DS, Rault L, Berkova N, Le Loir Y, Even S. Inhibition of *Staphylococcus aureus* invasion into bovine mammary epithelial cells by contact with live *Lactobacillus casei.* Appl Environ Microbiol 2013; 79(3): 877-85.
[http://dx.doi.org/10.1128/AEM.03323-12] [PMID: 23183972]

[32] Al-Qumber M, Tagg JR. Commensal bacilli inhibitory to mastitis pathogens isolated from the udder microbiota of healthy cows. J Appl Microbiol 2006; 101(5): 1152-60.
[http://dx.doi.org/10.1111/j.1365-2672.2006.03004.x] [PMID: 17040239]

[33] Otero MC, Nader-Macías ME. Inhibition of *Staphylococcus aureus* by H_2O_2-producing *Lactobacillus gasseri* isolated from the vaginal tract of cattle. Anim Reprod Sci 2006; 96(1-2): 35-46.
[http://dx.doi.org/10.1016/j.anireprosci.2005.11.004] [PMID: 16386389]

[34] Frola ID, Pellegrino MS, Espeche MC, Giraudo JA, Nader-Macias ME, Bogni CI. Effects of intramammary inoculation of *Lactobacillus perolens* CRL1724 in lactating cows' udders. J Dairy Res 2012; 79(1): 84-92.
[http://dx.doi.org/10.1017/S0022029911000835] [PMID: 22077995]

[35] Moreau MM, Eades SC, Reinemeyer CR, Fugaro MN, Onishi JC. Illumina sequencing of the V4 hypervariable region 16S rRNA gene reveals extensive changes in bacterial communities in the cecum following carbohydrate oral infusion and development of early-stage acute laminitis in the horse. Vet Microbiol 2014; 168(2-4): 436-41.
[http://dx.doi.org/10.1016/j.vetmic.2013.11.017] [PMID: 24355533]

[36] Costa MC, Arroyo LG, Allen-Vercoe E, *et al.* Comparison of the fecal microbiota of healthy horses and horses with colitis by high throughput sequencing of the V3-V5 region of the 16S rRNA gene. PLoS One 2012; 7(7): e41484.
[http://dx.doi.org/10.1371/journal.pone.0041484] [PMID: 22859989]

[37] Cebrián R, Baños A, Valdivia E, Pérez-Pulido R, Martínez-Bueno M, Maqueda M. Characterization of functional, safety, and probiotic properties of *Enterococcus faecalis* UGRA10, a new AS-48-producer strain. Food Microbiol 2012; 30(1): 59-67.

[http://dx.doi.org/10.1016/j.fm.2011.12.002] [PMID: 22265284]

[38] Mater DD, Langella P, Corthier G, Flores MJ. A probiotic *Lactobacillus* strain can acquire vancomycin resistance during digestive transit in mice. J Mol Microbiol Biotechnol 2008; 14(1-3): 123-7.
[http://dx.doi.org/10.1159/000106091] [PMID: 17957119]

[39] Desrochers AM, Dolente BA, Roy MF, Boston R, Carlisle S. Efficacy of *Saccharomyces boulardii* for treatment of horses with acute enterocolitis. J Am Vet Med Assoc 2005; 227(6): 954-9.
[http://dx.doi.org/10.2460/javma.2005.227.954] [PMID: 16190596]

[40] Boyle AG, Magdesian KG, Durando MM, Gallop R, Sigdel S. *Saccharomyces boulardii* viability and efficacy in horses with antimicrobial-induced diarrhoea. Vet Rec 2013; 172(5): 128.
[http://dx.doi.org/10.1136/vr.100833] [PMID: 23161811]

[41] Weese JS, Rousseau J. Evaluation of *Lactobacillus pentosus* WE7 for prevention of diarrhea in neonatal foals. J Am Vet Med Assoc 2005; 226(12): 2031-4.
[http://dx.doi.org/10.2460/javma.2005.226.2031] [PMID: 15989186]

[42] Yuyama T, Takai S, Tsubaki S, Kado Y, Morotomi M. Evaluation of a host-specific *Lactobacillus* probiotic in training-horses and neonatal foals. J Ind Microbiol 2004; 18: 101-6.

[43] Kenny M, Smidt H, Mengheri E, Miller B. Probiotics - do they have a role in the pig industry? Animal 2011; 5(3): 462-70.
[http://dx.doi.org/10.1017/S175173111000193X] [PMID: 22445413]

[44] Böhmer BM, Kramer W, Roth-Maier DA. Dietary probiotic supplementation and resulting effects on performance, health status, and microbial characteristics of primiparous sows. J Anim Physiol Anim Nutr (Berl) 2006; 90(7-8): 309-15.
[http://dx.doi.org/10.1111/j.1439-0396.2005.00601.x] [PMID: 16867076]

[45] Taras D, Vahjen W, Macha M, Simon O. Response of performance characteristics and fecal consistency to long-lasting dietary supplementation with the probiotic strain *Bacillus cereus* var. *toyoi* to sows and piglets. Arch Anim Nutr 2005; 59(6): 405-17.
[http://dx.doi.org/10.1080/17450390500353168] [PMID: 16429826]

[46] Heo JM, Opapeju FO, Pluske JR, Kim JC, Hampson DJ, Nyachoti CM. Gastrointestinal health and function in weaned pigs: a review of feeding strategies to control post-weaning diarrhoea without using in-feed antimicrobial compounds. J Anim Physiol Anim Nutr (Berl) 2013; 97(2): 207-37.
[http://dx.doi.org/10.1111/j.1439-0396.2012.01284.x] [PMID: 22416941]

[47] Siggers RH, Siggers J, Boye M, *et al.* Early administration of probiotics alters bacterial colonization and limits diet-induced gut dysfunction and severity of necrotizing enterocolitis in preterm pigs. J Nutr 2008; 138(8): 1437-44.
[PMID: 18641188]

[48] Konstantinov SR, Smidt H, Akkermans AD, *et al.* Feeding of *Lactobacillus sobrius* reduces *Escherichia coli* F4 levels in the gut and promotes growth of infected piglets. FEMS Microbiol Ecol 2008; 66(3): 599-607.
[http://dx.doi.org/10.1111/j.1574-6941.2008.00517.x] [PMID: 18537838]

[49] Pieper R, Janczyk P, Urubschurov V, Korn U, Pieper B, Souffrant WB. Effect of a single oral

administration of *Lactobacillus plantarum* DSMZ 8862/8866 before and at the time point of weaning on intestinal microbial communities in piglets. Int J Food Microbiol 2009; 130(3): 227-32. [http://dx.doi.org/10.1016/j.ijfoodmicro.2009.01.026] [PMID: 19237219]

[50] Bontempo V, Di Giancamillo A, Savoini G, Dell'Orto V, Domeneghini C. Live yeast dietary supplementation acts upon intestinal morpho-functional aspects and growth in weanling piglets. Anim Feed Sci Technol 2006; 129: 224-36. [http://dx.doi.org/10.1016/j.anifeedsci.2005.12.015]

[51] Kesarcodi-Watson A, Kaspar H, Lategan MJ, Gibson L. Probiotics in aquaculture: The need, principles and mechanisms of action and screening processes. Aquaculture 2008; 274: 1-14. [http://dx.doi.org/10.1016/j.aquaculture.2007.11.019]

[52] Bergh Ø, Naas KE, Harboe T. Shift in the intestinal microflora of atlantic halibut (*Hippoglossus hippoglossus*) larvae during first feeding. Can J Fish Aquat Sci 1994; 51: 1899-903. [http://dx.doi.org/10.1139/f94-190]

[53] Jorquera MA, Silva FR, Riquelme CE. Bacteria in the culture of the scallop *Argopecten purpuratus* (Lamarck, 1819). Aquacult Int 2001; 9: 285-303. [http://dx.doi.org/10.1023/A:1020449324456]

[54] Newaj-Fyzul A, Austin B. Probiotics, immunostimulants, plant products and oral vaccines, and their role as feed supplements in the control of bacterial fish diseases. J Fish Dis 2015; 38(11): 937-55. [http://dx.doi.org/10.1111/jfd.12313] [PMID: 25287254]

[55] Das S, Ward LR, Burke C. Prospects of using marine actinobacteria as probiotics in aquaculture. Appl Microbiol Biotechnol 2008; 81(3): 419-29. [http://dx.doi.org/10.1007/s00253-008-1731-8] [PMID: 18841358]

[56] Lara-Flores M, Olvera-Novoa MA. The use of lactic acid bacteria isolated from intestinal tract of Nile tilapia (*Oreochromis niloticus*), as growth promoters in fish fed low protein diets. Lat Am J Aquat Res 2013; 41: 490-7.

[57] Lobo C, Moreno-Ventas X, Tapia-Paniagua S, Rodríguez C, Moriñigo MA, de La Banda IG. Dietary probiotic supplementation (*Shewanella putrefaciens* Pdp11) modulates gut microbiota and promotes growth and condition in Senegalese sole larviculture. Fish Physiol Biochem 2014; 40(1): 295-309. [http://dx.doi.org/10.1007/s10695-013-9844-0] [PMID: 23933744]

[58] Ramos MA, Weber B, Gonçalves JF, Santos GA, Rema P, Ozório RO. Dietary probiotic supplementation modulated gut microbiota and improved growth of juvenile rainbow trout (*Oncorhynchus mykiss*). Comp Biochem Physiol A Mol Integr Physiol 2013; 166(2): 302-7. [http://dx.doi.org/10.1016/j.cbpa.2013.06.025] [PMID: 23811044]

[59] Xing C-F, Hu H-H, Huang J-B, *et al.* Diet supplementation of *Pediococcus pentosaceus* in cobia (*Rachycentron canadum*) enhances growth rate, respiratory burst and resistance against photobacteriosis. Fish Shellfish Immunol 2013; 35(4): 1122-8. [http://dx.doi.org/10.1016/j.fsi.2013.07.021] [PMID: 23916590]

[60] Antony SP, Singh IS, Jose RM, Kumar PR, Philip R. Antimicrobial peptide gene expression in tiger shrimp, *Penaeus monodon* in response to gram-positive bacterial probionts and white spot virus challenge. Aquaculture 2011; 316: 6-12. [http://dx.doi.org/10.1016/j.aquaculture.2011.03.025]

[61] Balcázar JL, de Blas I, Ruiz-Zarzuela I, Cunningham D, Vendrell D, Múzquiz JL. The role of probiotics in aquaculture. Vet Microbiol 2006; 114(3-4): 173-86. [http://dx.doi.org/10.1016/j.vetmic.2006.01.009] [PMID: 16490324]

[62] Das A, Nakhro K, Chowdhury S, Kamilya D. Effects of potential probiotic *Bacillus amyloliquefaciens* FPTB16 on systemic and cutaneous mucosal immune responses and disease resistance of catla *(Catla catla)*. Fish Shellfish Immunol 2013; 35(5): 1547-53. [http://dx.doi.org/10.1016/j.fsi.2013.08.022] [PMID: 24012512]

[63] Castex M, Chim L, Pham D, *et al.* Probiotic P-acidilactici application in shrimp *Litopenaeus stylirostris* culture subject to vibriosis in New Caledonia. Aquaculture 2008; 31(275): 182-93. [http://dx.doi.org/10.1016/j.aquaculture.2008.01.011]

[64] Huang J-B, Wu Y-C, Chi S-C. Dietary supplementation of *Pediococcus pentosaceus* enhances innate immunity, physiological health and resistance to *Vibrio anguillarum* in orange-spotted grouper (*Epinephelus coioides*). Fish Shellfish Immunol 2014; 39(2): 196-205. [http://dx.doi.org/10.1016/j.fsi.2014.05.003] [PMID: 24845519]

[65] Kim D-H, Austin B. Cytokine expression in leucocytes and gut cells of rainbow trout, *Oncorhynchus mykiss* Walbaum, induced by probiotics. Vet Immunol Immunopathol 2006; 114(3-4): 297-304. [http://dx.doi.org/10.1016/j.vetimm.2006.08.015] [PMID: 17011045]

[66] Nikoskelainen S, Ouwehand AC, Bylund G, Salminen S, Lilius E-M. Immune enhancement in rainbow trout (*Oncorhynchus mykiss*) by potential probiotic bacteria (*Lactobacillus rhamnosus*). Fish Shellfish Immunol 2003; 15(5): 443-52. [http://dx.doi.org/10.1016/S1050-4648(03)00023-8] [PMID: 14550670]

[67] Irianto A, Austin B. Use of dead probiotic cells to control furunculosis in rainbow trout, *Oncorhynchus mykiss* (Walbaum). J Fish Dis 2003; 26(1): 59-62. [http://dx.doi.org/10.1046/j.1365-2761.2003.00414.x] [PMID: 12962213]

[68] Marshall-Jones ZV, Baillon M-L, Croft JM, Butterwick RF. Effects of *Lactobacillus acidophilus* DSM13241 as a probiotic in healthy adult cats. Am J Vet Res 2006; 67(6): 1005-12. [http://dx.doi.org/10.2460/ajvr.67.6.1005] [PMID: 16740094]

[69] Veir JK, Knorr R, Cavadini C, *et al.* Effect of supplementation with *Enterococcus faecium* (SF68) on immune functions in cats. Vet Ther 2007; 8(4): 229-38. [PMID: 18183541]

[70] Lappin MR, Veir JK, Satyaraj E, Czarnecki-Maulden G. Pilot study to evaluate the effect of oral supplementation of *Enterococcus faecium* SF68 on cats with latent feline herpesvirus 1. J Feline Med Surg 2009; 11(8): 650-4. [http://dx.doi.org/10.1016/j.jfms.2008.12.006] [PMID: 19201238]

[71] Benyacoub J, Czarnecki-Maulden GL, Cavadini C, *et al.* Supplementation of food with *Enterococcus faecium* (SF68) stimulates immune functions in young dogs. J Nutr 2003; 133(4): 1158-62. [PMID: 12672936]

[72] Rinkinen M, Jalava K, Westermarck E, Salminen S, Ouwehand AC. Interaction between probiotic lactic acid bacteria and canine enteric pathogens: a risk factor for intestinal *Enterococcus faecium* colonization? Vet Microbiol 2003; 92(1-2): 111-9.

[http://dx.doi.org/10.1016/S0378-1135(02)00356-5] [PMID: 12488075]

[73] O'Mahony D, Murphy KB, MacSharry J, *et al.* Portrait of a canine probiotic *Bifidobacterium*--from gut to gut. Vet Microbiol 2009; 139(1-2): 106-12. [http://dx.doi.org/10.1016/j.vetmic.2009.05.002] [PMID: 19523775]

[74] Sauter SN, Allenspach K, Gaschen F, Gröne A, Ontsouka E, Blum JW. Cytokine expression in an *ex vivo* culture system of duodenal samples from dogs with chronic enteropathies: modulation by probiotic bacteria. Domest Anim Endocrinol 2005; 29(4): 605-22. [http://dx.doi.org/10.1016/j.domaniend.2005.04.006] [PMID: 15941645]

[75] Schmitz S, Henrich M, Neiger R, Werling D, Allenspach K. Stimulation of duodenal biopsies and whole blood from dogs with food-responsive chronic enteropathy and healthy dogs with Toll-like receptor ligands and probiotic *Enterococcus faecium.* Scand J Immunol 2014; 80(2): 85-94. [http://dx.doi.org/10.1111/sji.12186] [PMID: 24813376]

[76] Rossi G, Pengo G, Caldin M, *et al.* Comparison of microbiological, histological, and immunomodulatory parameters in response to treatment with either combination therapy with prednisone and metronidazole or probiotic VSL#3 strains in dogs with idiopathic inflammatory bowel disease. PLoS One 2014; 9(4): e94699. [http://dx.doi.org/10.1371/journal.pone.0094699] [PMID: 24722235]

[77] Stone R. Bacteriophage therapy. Stalin's forgotten cure. Science 2002; 298(5594): 728-31. [http://dx.doi.org/10.1126/science.298.5594.728] [PMID: 12399562]

[78] Chatain-Ly MH. The factors affecting effectiveness of treatment in phages therapy, mini review. Front Microbiol 2014; 18: 5.

[79] Sulakvelidze A. Safety by nature: potential bacteriophage applications. Microbes 2011; 6: 122-6.

[80] Jassim SA, Limoges R. Natural solution to antibiotic resistance: bacteriophages 'The Living Drugs'. World J Microbiol Biotechnol 2014 2014; 30(8): 2153-70. 2014/08/01

[81] Edgar R, Friedman N, Molshanski-Mor S, Qimron U. Reversing bacterial resistance to antibiotics by phage-mediated delivery of dominant sensitive genes. Appl Environ Microbiol 2011. [http://dx.doi.org/10.1128/AEM.05741-11] [PMID: 22113912]

[82] Bren L. Bacteria-eating virus approved as food additive. FDA Consum 2007; 41(1): 20-2. [PMID: 17342833]

[83] Fiorentin L, Vieira ND, Barioni W Jr. Oral treatment with bacteriophages reduces the concentration of *Salmonella* Enteritidis PT4 in caecal contents of broilers. Avian Pathol 2005; 34(3): 258-63. [http://dx.doi.org/10.1080/01445340500112157] [PMID: 16191711]

[84] Lim T-H, Kim M-S, Lee D-H, *et al.* Use of bacteriophage for biological control of *Salmonella* Enteritidis infection in chicken. Res Vet Sci 2012; 93(3): 1173-8. [http://dx.doi.org/10.1016/j.rvsc.2012.06.004] [PMID: 22795674]

[85] Henriques A, Sereno R, Almeida A. Reducing *Salmonella* horizontal transmission during egg incubation by phage therapy. Foodborne Pathog Dis 2013; 10(8): 718-22. [http://dx.doi.org/10.1089/fpd.2012.1363] [PMID: 23672473]

[86] Wagenaar JA, Van Bergen MA, Mueller MA, Wassenaar TM, Carlton RM. Phage therapy reduces *Campylobacter jejuni* colonization in broilers. Vet Microbiol 2005; 109(3-4): 275-83.

[http://dx.doi.org/10.1016/j.vetmic.2005.06.002] [PMID: 16024187]

[87] Carvalho CM, Gannon BW, Halfhide DE, *et al.* The *in vivo* efficacy of two administration routes of a phage cocktail to reduce numbers of *Campylobacter coli* and *Campylobacter jejuni* in chickens. BMC Microbiol 2010; 10: 232.
[http://dx.doi.org/10.1186/1471-2180-10-232] [PMID: 20809975]

[88] Kittler S, Fischer S, Abdulmawjood A, Glünder G, Klein G. Effect of bacteriophage application on *Campylobacter jejuni* loads in commercial broiler flocks. Appl Environ Microbiol 2013; 79(23): 7525-33.
[http://dx.doi.org/10.1128/AEM.02703-13] [PMID: 24077703]

[89] Bigot B, Lee WJ, McIntyre L, *et al.* Control of *Listeria monocytogenes* growth in a ready-to-eat poultry product using a bacteriophage. Food Microbiol 2011; 28(8): 1448-52.
[http://dx.doi.org/10.1016/j.fm.2011.07.001] [PMID: 21925027]

[90] Oliveira M, Viñas I, Colàs P, Anguera M, Usall J, Abadias M. Effectiveness of a bacteriophage in reducing *Listeria monocytogenes* on fresh-cut fruits and fruit juices. Food Microbiol 2014; 38: 137-42.
[http://dx.doi.org/10.1016/j.fm.2013.08.018] [PMID: 24290636]

[91] Jun JW, Kim HJ, Yun SK, Chai JY, Park SC. Eating oysters without risk of vibriosis: application of a bacteriophage against *Vibrio parahaemolyticus* in oysters. Int J Food Microbiol 2014; 188: 31-5.
[http://dx.doi.org/10.1016/j.ijfoodmicro.2014.07.007] [PMID: 25086350]

[92] Richards GP. Bacteriophage remediation of bacterial pathogens in aquaculture: a review of the technology. Bacteriophage 2014; 4: e975540.
[http://dx.doi.org/10.4161/21597081.2014.975540]

[93] Khairnar K, Raut MP, Chandekar RH, Sanmukh SG, Paunikar WN. Novel bacteriophage therapy for controlling metallo-beta-lactamase producing *Pseudomonas aeruginosa* infection in catfish. BMC Vet Res 2013; 9: 264-73.
[http://dx.doi.org/10.1186/1746-6148-9-264] [PMID: 24369750]

[94] Park SC, Nakai T. Bacteriophage control of *Pseudomonas plecoglossicida* infection in ayu *Plecoglossus altivelis.* Dis Aquat Organ 2003; 53(1): 33-9.
[http://dx.doi.org/10.3354/dao053033] [PMID: 12608566]

[95] Jun JW, Kim JH, Shin SP, Han JE, Chai JY, Park SC. Protective effects of the *Aeromonas* phages pAh1-C and pAh6-C against mass mortality of the cyprinid loach (*Misgurnus anguillicaudatus*) caused by *Aeromonas hydrophila.* Aquaculture 2013; 416: 289-95.
[http://dx.doi.org/10.1016/j.aquaculture.2013.09.045]

[96] Madsen L, Bertelsen SK, Dalsgaard I, Middelboe M. Dispersal and survival of *Flavobacterium psychrophilum* phages *in vivo* in rainbow trout and *in vitro* under laboratory conditions: implications for their use in phage therapy. Appl Environ Microbiol 2013; 79(16): 4853-61.
[http://dx.doi.org/10.1128/AEM.00509-13] [PMID: 23747702]

[97] Zhang C, Li W, Liu W, *et al.* T4-like phage Bp7, a potential antimicrobial agent for controlling drug-resistant *Escherichia coli* in chickens. Appl Environ Microbiol 2013; 79(18): 5559-65.
[http://dx.doi.org/10.1128/AEM.01505-13] [PMID: 23835183]

[98] Tsonos J, Oosterik LH, Tuntufye HN, *et al.* A cocktail of *in vitro* efficient phages is not a guarantee

for *in vivo* therapeutic results against avian colibacillosis. Vet Microbiol 2014; 171(3-4): 470-9. [http://dx.doi.org/10.1016/j.vetmic.2013.10.021] [PMID: 24269008]

[99] Atterbury RJ, Van Bergen MA, Ortiz F, *et al.* Bacteriophage therapy to reduce *salmonella* colonization of broiler chickens. Appl Environ Microbiol 2007; 73(14): 4543-9. [http://dx.doi.org/10.1128/AEM.00049-07] [PMID: 17526794]

[100] Bardina C, Spricigo DA, Cortés P, Llagostera M. Significance of the bacteriophage treatment schedule in reducing *Salmonella* colonization of poultry. Appl Environ Microbiol 2012; 78(18): 6600-7. [http://dx.doi.org/10.1128/AEM.01257-12] [PMID: 22773654]

[101] Loc Carrillo C, Atterbury RJ, el-Shibiny A, *et al.* Bacteriophage therapy to reduce *Campylobacter jejuni* colonization of broiler chickens. Appl Environ Microbiol 2005; 71(11): 6554-63. [http://dx.doi.org/10.1128/AEM.71.11.6554-6563.2005] [PMID: 16269681]

[102] Dias RS, Eller MR, Duarte VS, *et al.* Use of phages against antibiotic-resistant *Staphylococcus aureus* isolated from bovine mastitis. J Anim Sci 2013; 91(8): 3930-9. [http://dx.doi.org/10.2527/jas.2012-5884] [PMID: 23893999]

[103] Li L, Zhang Z. Isolation and characterization of a virulent bacteriophage SPW specific for *Staphylococcus aureus* isolated from bovine mastitis of lactating dairy cattle. Mol Biol Rep 2014; 41(9): 5829-38. [http://dx.doi.org/10.1007/s11033-014-3457-2] [PMID: 24981924]

[104] Kwiatek M, Parasion S, Mizak L, Gryko R, Bartoszcze M, Kocik J. Characterization of a bacteriophage, isolated from a cow with mastitis, that is lytic against *Staphylococcus aureus* strains. Arch Virol 2012; 157(2): 225-34. [http://dx.doi.org/10.1007/s00705-011-1160-3] [PMID: 22045271]

[105] Endersen L, Coffey A, Neve H, McAuliffe O, Ross RP, O'Mahony JM. Isolation and characterisation of six novel mycobacteriophages and investigation of their antimicrobial potential in milk. Int Dairy J 2013; 28: 8-14. [http://dx.doi.org/10.1016/j.idairyj.2012.07.010]

[106] Callaway TR, Edrington TS, Brabban AD, *et al.* Bacteriophage isolated from feedlot cattle can reduce *Escherichia coli* O157:H7 populations in ruminant gastrointestinal tracts. Foodborne Pathog Dis 2008; 5(2): 183-91. [http://dx.doi.org/10.1089/fpd.2007.0057] [PMID: 18407757]

[107] Bach SJ, Johnson RP, Stanford K, McAllister TA. Bacteriophages reduce *Escherichia coli* O157:H7 levels in experimentally inoculated sheep. Can J Anim Sci 2009; 89: 285-93. [http://dx.doi.org/10.4141/CJAS08083]

[108] Saez AC, Zhang J, Rostagno MH, Ebner PD. Direct feeding of microencapsulated bacteriophages to reduce *Salmonella* colonization in pigs. Foodborne Pathog Dis 2011; 8(12): 1269-74. [http://dx.doi.org/10.1089/fpd.2011.0905] [PMID: 21854261]

[109] Callaway TR, Edrington TS, Brabban A, *et al.* Evaluation of phage treatment as a strategy to reduce *Salmonella* populations in growing swine. Foodborne Pathog Dis 2011; 8(2): 261-6. [http://dx.doi.org/10.1089/fpd.2010.0671] [PMID: 21034249]

[110] Hooton SP, Atterbury RJ, Connerton IF. Application of a bacteriophage cocktail to reduce *Salmonella*

Typhimurium U288 contamination on pig skin. Int J Food Microbiol 2011; 151(2): 157-63. [http://dx.doi.org/10.1016/j.ijfoodmicro.2011.08.015] [PMID: 21899907]

[111] Pirnay J-P, Verbeken G, Rose T, *et al.* Introducing yesterday's phage therapy in today's medicine. Future Virol 2012; 7: 379-90. [http://dx.doi.org/10.2217/fvl.12.24]

[112] Tsonos J, Vandenheuvel D, Briers Y, De Greve H, Hernalsteens J-P, Lavigne R. Hurdles in bacteriophage therapy: deconstructing the parameters. Vet Microbiol 2014; 171(3-4): 460-9. [http://dx.doi.org/10.1016/j.vetmic.2013.11.001] [PMID: 24315040]

[113] Jamalludeen N, Johnson RP, Shewen PE, Gyles CL. Evaluation of bacteriophages for prevention and treatment of diarrhea due to experimental enterotoxigenic *Escherichia coli* O149 infection of pigs. Vet Microbiol 2009; 136(1-2): 135-41. [http://dx.doi.org/10.1016/j.vetmic.2008.10.021] [PMID: 19058927]

[114] Dini C, Islan GA, de Urraza PJ, Castro GR. Novel biopolymer matrices for microencapsulation of phages: enhanced protection against acidity and protease activity. Macromol Biosci 2012; 12(9): 1200-8. [http://dx.doi.org/10.1002/mabi.201200109] [PMID: 22847825]

[115] Guenther S, Huwyler D, Richard S, Loessner MJ. Virulent bacteriophage for efficient biocontrol of *Listeria monocytogenes* in ready-to-eat foods. Appl Environ Microbiol 2009; 75(1): 93-100. [http://dx.doi.org/10.1128/AEM.01711-08] [PMID: 19011076]

[116] Gill JJ, Sabour PM, Leslie KE, Griffiths MW. Bovine whey proteins inhibit the interaction of *Staphylococcus aureus* and bacteriophage K. J Appl Microbiol 2006; 101(2): 377-86. [http://dx.doi.org/10.1111/j.1365-2672.2006.02918.x] [PMID: 16882145]

Summary

Frontiers in Antimicrobial Agents –The Challenging of Antibiotic Resistance in the Development of New Therapeutics

The prevalence of drug-resistant microorganisms has increased worldwide [1]. However, in the last years, few novel entities to fight drug resistant microorganisms have entered the clinic. The antibacterial pipeline is scarce because of the costs associated with the clinical trials and licensing of the antibiotics [2]. The emergence of multiresistant strains and the temporal and geographical variations in resistance patterns suggest the need to look for alternatives to the classic antimicrobial therapy. The development of novel entities to control the dissemination of drug resistant microorganisms would decrease the morbidity and mortality caused by these disease causing agents, and consequently the economic burden associated with health treatments.

The present book, entitled "Frontiers in Antimicrobial Agents – The challenging of antibiotic resistance in the development of new therapeutics", gathers a wide range of topics on the subject and includes several chapters with original material. Authors and co-authors represent a multidisciplinary team that includes scientists with a vast experience in the area, from different universities and research institutions. Topics addressed include: Phage therapy, Antimicrobial Peptides (AMPs), Probiotics, Immunotherapy, Natural Compounds, Bacteriocins, Biocides, and Novel Therapeutic Strategies in Veterinary Medicine. The book describes innovative alternatives to the classical antimicrobial therapeutics, based on cutting edge research, representing a huge advance to science and technology. Some of these approaches are considered extremely valuable in the nearly future due to their low toxicity, capacity for large scale production, and most importantly their low probability to generate resistance.

The book will be extremely valuable for all those who are interested in antibiotic resistance, and in strategies to the classic antibiotherapy being a significant advance to the state-of-the-art in this regard. The variety of areas focused will attract a broad range of readers, including researchers, clinicians, professors and students. It will also enlighten readers regarding different approaches to control the dissemination of drug resistant microorganisms.

[1]. World Health Organization (WHO). Antimicrobial Resistance - Global Report on

Surveillance. April 2014.

[2]. Gupta SK, Nayak RP. Dry antibiotic pipeline: Regulatory bottlenecks and regulatory reforms. J Pharmacol Pharmacother. 2014 Jan;5(1):4-7. doi: 10.4103/0976-500X.124405.

SUBJECT INDEX

S

T

V

Made in the USA
Las Vegas, NV
03 November 2026

57424577R00169